Uncertain Science... Uncertain

Why can't science answer, once and for all, the major questions that make up our headlines:
Is the world warming because of the greenhouse effect?
What would be the dangers associated with a terrorist release of anthrax spores?
What action should be taken against an outbreak of foot-and-mouth disease or BSE?
Why can't we predict the occurrences of earthquakes?

Scientific uncertainty puzzles many people. The puzzlement arises when scientists have more than one answer and disagree among themselves. *Uncertain Science... Uncertain World* will help people to find their way through a maze of contradiction and uncertainty. By acquainting them with the ways that uncertainty arises in science, how scientists accommodate and make use of uncertainty, and how they reach conclusions in the face of uncertainty, the book will enable the reader confidently to evaluate uncertainty from their own perspectives, in terms of their own everyday experiences.

Advance praise for *Uncertain Science... Uncertain World*

'*Uncertain Science... Uncertain World* gives the layman an excellent inside look at how science works and flourishes even though it is immersed in uncertainty. Pollack analyses the paradox that society is unable or unwilling to address environmental problems of global scale – often under the pretence that there's not enough scientific certainty to take action – while at the same time the insurance industry and other businesses routinely hedge the risks attendant to an uncertain future. It's my hope that this very clearly written book, devoid of both polemics and equations, will be widely read by the general public and policy-makers.'
Paul Crutzen, Winner of the 1995 Nobel Prize for Chemistry for work on the ozone hole

'*Uncertain Science... Uncertain World* is certain to clarify one of the most fundamental popular misconceptions about science – that it is exact and

certain. Henry Pollack demolishes the mythology about certainty in science with short and clear examples of how uncertainty is both endemic to science and not a cause for paralysis or inaction. This well-written book is a welcome antidote to the misrepresentations of special interests, who misuse scientific uncertainty to stall public policy and advance their own agendas.'
Stephen Schneider, Professor of Environmental Biology at Stanford University and author of Laboratory Earth: The Planetary Gamble We Can't Afford to Lose

'This excellent book will serve as a blast of common sense to counter two dangerous attitudes. One is the desperate search for impossible certainties in a complex world where few comprehend the meaning of probability. The other is a belief that scientists are the magicians of today who can deliver certainty by 'scientific tests'. Pollack writes with vigour and clarity about big issues such as global warming, and reading this book ought to help us to become better judges when 'facts' conflict. There are few more important attributes we need for the twenty-first century.'
Aubrey Manning, Professor Emeritus at the University of Edinburgh and author of An Introduction to Animal Behaviour

'Public policy debates are constantly getting stuck in the mire of perceptions about scientific uncertainty and risk. Yet science is no different to many other areas of human experience in that uncertainty and risk are inevitably present. In a readable, entertaining presentation, Henry Pollack removes some of the mystery surrounding scientific uncertainty by placing it alongside examples from everyday life.'
Sir John Houghton, Co-chair of the Intergovernmental Panel on Climate Change and author of Global Warming – The Complete Briefing

'At last we have a solid, scientific look at the vexing subject of uncertainty. You may not be more certain about some subjects when you finish this book, but you'll understand why.'
James Trefil, Professor of Physics at George Mason University, and author of A Scientist in the City

'Too often, scientists fall into the quicksand of technical jargon and fail to communicate important information to the general public. In *Uncertain Science...Uncertain World*, Henry Pollack uses plain English and engaging examples to explore uncertainty both in science and everyday life.'
Neal Lane, Professor at Rice University, former Science Advisor to President Clinton and former Director of the US National Science Foundation

Uncertain Science...
Uncertain World

HENRY N. POLLACK

CAMBRIDGE
UNIVERSITY PRESS

CAMBRIDGE UNIVERSITY PRESS
Cambridge, New York, Melbourne, Madrid, Cape Town, Singapore, São Paulo

Cambridge University Press
The Edinburgh Building, Cambridge CB2 2RU, UK
Published in the United States of America by Cambridge University Press,
New York

www.cambridge.org
Information on this title: www.cambridge.org/9780521781884

First published in hardback 2003
Paperback edition first published 2005

Printed in the United Kingdom at the University Press, Cambridge

A catalogue record for this book is available from the British Library

Library of Congress Cataloging in Publication data

Pollack, H. N.
Uncertain Science ... Uncertain World / Henry N. Pollack.
    p. cm.
Includes bibliographical references and index.
ISBN 0 521 78188 4

1. Science – Philosophy.   2. Uncertainty.   I. Title.
Q175 .P835   2003
501 – dc21   2002031200

ISBN-13 978-0-521-78188-4 hardback
ISBN-10 0-521-78188-4 hardback

ISBN-13 978-0-521-61910-6 paperback
ISBN-10 0-521-61910-6 paperback

To Lana, John and Sara...the loves of my life

Contents

Acknowledgments

I am very grateful to my wife Lana and son John for reading the manuscript at several stages of development, and offering their sometimes harsh but always insightful comments and suggestions. Both are frequent and excellent writers in their own right, and I have benefited from having such capable editors close to home. Jason Smerdon, Boris Kiefer, David Chapman, Drew Isaacs, and Gordon Kane have also read all or parts of the manuscript and helped it along in many ways. Kooiti Masuda kindly called to my attention a few errors of fact that appeared in the first hardback printing of the book. I also thank Matt Lloyd, my editor at Cambridge University Press, for early encouragement and later critical commentary. Needless to say, all of these helpful readers bear no responsibility for errors and pointed opinions.

About the author

No book can be free of the background and experiences of the author, so let me tell you a little about myself.

I was born and schooled in Nebraska, in the agricultural heartland of America. My mother was a traditional homemaker, my father raised livestock on the family farm. As a young boy in Nebraska, I thought the world was made of dirt, good rich soil that with a lot of hard work yielded good things to eat. In 1954, at age eighteen, I went off to college at Cornell University in upstate New York, a long way from my midlands home. At Cornell, the bedrock is well exposed in deep gorges carved by small streams tumbling down to Cayuga Lake, one of the spectacular glacially sculpted Finger Lakes of New York. There I learned that the soil was only a thin veneer on top of layers and layers of rock, the real *terra firma*. And in those layers were fossils, the record of life on Earth in ages past. I was awestruck by the vastness of time so revealed, and in virtually no time I was firmly hooked on geology as a career choice. The entire Earth was my field of study, and in a sense I never returned home: "How can you keep 'em down on the farm, after they've seen the Paleozoic?"

I did return to the University of Nebraska for a master's degree (lured by an excellent faculty in geology and the bargain tuition of $90 per semester) and then went on to the University of Michigan to study for a PhD and to Harvard for a postdoctoral research position. Between Michigan and Harvard, I married Lana Schoenberger, a Michigan girl, and when a teaching position was advertised at the University of Michigan a few years later, I interviewed and was delighted to be selected for the job. Lana and I have been in Ann Arbor, home of the University of Michigan, ever since. We had two children, Sara whose life was cut short by an accident at age 14, and John who is an author and writer living in Washington, DC. As a family we twice lived abroad, in Zambia in 1970–71, and in England in 1977–78.

At the University of Michigan, I have taught at every level of the curriculum, from introductory undergraduate Earth Science courses for non-scientists to specialized graduate seminars, and I have taught in virtually every setting: lecture, laboratory, seminar, and in the field. In my department, unlike some others, responsibility for introductory courses is placed in the hands of the more experienced faculty, and consequently I have increasingly taught courses for undergraduates who will not be pursuing scientific careers. The challenge in such courses is to develop in students an awareness of the ways science interfaces with their lives, and to enable them to understand both the strengths and frailties of the 'scientific method'. I have taught the generic first course in geology, 'Geology 101', and other courses with titles like Geology of the National Parks, Climate and Mankind, Geology and Climate of the Planets, and Science and Politics of Global Warming.

My principal research efforts for many years addressed Earth's internal heat and how that heat is lost over time. The Earth's heat is the fuel for the big 'engine' that drives plate tectonics and continental drift, yielding earthquakes and volcanoes as a byproduct. Those large-scale processes that shape Earth's surface are actually manifestations of the internal processes that enable the planet to cool slowly. For many years, my students and I made field measurements of the heat loss from the Earth's interior in Africa, South America and in the USA. In simple terms, when people ask me to describe that process, I often reply that I go out and take the Earth's temperature in wonderfully remote places around the globe.

Over the past decade, my geothermal research has taken a new direction. My colleagues and I came to recognize that the temperature profiles of the outer few thousand feet of the Earth's crust comprised an archive of the planet's changing climate over the past millennium. The underlying principle is that if the surface of the Earth is warming (or cooling for that matter) the rocks below the surface will feel it and record it. The longer the change at the surface persists, the deeper the warming will penetrate into the subsurface. In the context of the important debate about global warming and its probable causes, the thermal information contained at these depths in the Earth's crust enables a comparison of the Earth's surface temperature in both the industrial and pre-industrial era, and it provides an estimate of the size of the human contribution to climate change.

In an academic career of this duration, I have also had the opportunity to gain administrative experience. I led the Department of Geological Sciences as Chairman for a period, and I also served the college administration as the Associate Dean for Research. In the national scientific community, I have had the opportunity to serve on several advisory panels for the US National Science Foundation's Division of Earth Sciences. These panels evaluate research proposals from scientists around the country who are seeking funding for their research programs. I also served four years on the American Geophysical Union's Committee on Global and Environmental Change, which among other tasks has been responsible for preparing the position statement of this professional organization on the subject of global climate change.

As is the custom at many universities, the faculty, in addition to their primary responsibilities in teaching and research, engage in service activities for the university, community, state, and nation. In my department, I spearheaded the development of our alumni relations program. Aside from the obvious benefits that financial contributions from alumni bring to the students and faculty of the department, there are the more subtle benefits that come from regular interactions with graduates out in the workaday world, an awareness of circumstances and constraints, problems and solutions that these people face in their professional lives, and their perceptions of science-based issues such as water quality standards, environmental cleanups, and global climate change. These interactions help academics to assess the relevance and effectiveness of their curricula and programs.

Over the past decade, I have frequently had the opportunity to discuss scientific issues such as global warming, radioactive waste disposal, and earthquake prediction with many groups outside the University. These discussions are generally with mature, thoughtful, educated people with non-scientific backgrounds. The venues for these discussions are diverse, but include meetings with University of Michigan alumni groups around the country and the world, talks to community service organizations such as Rotary and Kiwanis, workshops with professional journalists, expeditions to Antarctica with groups of eco-tourists, testimony before a US Senate committee, briefings at the White House, seminars for legislative staff at the state and federal level, interviews with the press and on radio and television, and call-in shows on public radio. Through these diverse activities, I have learned of the many misconceptions that educated people from all walks of life harbor about the scientific enterprise generally. Perhaps at the head of the list of misconceptions is the concept of uncertainty.

Scientific uncertainty puzzles many people, not because they have a hard time accepting that scientists do not have answers to every pressing question. The puzzlement arises when scientists have more than one answer and disagree among themselves. I have found that one of the principal contributions that I make as a teacher is to help people to find their way through a maze of contradiction and uncertainty. By acquainting them with the ways that uncertainty arises in science, how scientists accommodate and make use of uncertainty, and how they reach conclusions in the face of uncertainty, I have enabled them to evaluate uncertainty confidently from their own perspectives, in terms of their own experiences. This book has developed because I wish to share these views more widely.

Henry N. Pollack
Ann Arbor, Michigan
April 2002

I Setting the stage

This is a book about uncertainty, particularly the uncertainty we associate with science. Over the years, scientific uncertainty has been addressed by natural scientists, engineers, medical researchers, social scientists, and philosophers. But for all the perspectives that have been laid out in everything from short essays to scholarly monographs, the richness of scientific uncertainty has often been unappreciated and/or misunderstood by the general public, people not regularly engaged in science.

Uncertainty, of course, is not confined to the world of science. It is an everyday fact of ordinary life as well. We regularly face uncertainty in a myriad of ways. Will it rain today? Will Aunt Dorothy's plane arrive on time? Will the stock market tumble? Will an accident snarl the freeway during rush hour? These day-to-day uncertainties come and go, and we move on through life, sometimes preparing for them, but more often just plowing through them.

But uncertainty also colors longer-term concerns. Will my pension program be sufficient two decades from now to enable the full and comfortable life that my wife and I hope for? Will our health allow a free and independent life-style thirty years in the future? These longer-term questions are harder to answer and are cloaked in greater uncertainty. Because we have only one life to live we cannot return to 'Go' and take another path. Of necessity, we must plan, make decisions, and do our best, all the while evaluating our actions and making mid-course corrections according to our best judgment at the time.

Uncertainty is hardly confined to the future alone; it characterizes our knowledge of the past as well. Adopted children wonder about their birth-parentage, families have difficulty reconstructing the circumstances that led great-grandparents to emigrate. Military historians continue to reconstruct various scenarios for General Gordon's last days in Khartoum, or for Major Custer's last stand in the hills overlooking the Little Bighorn. Geologists are far from settled about the causes of ice ages, and paleontologists still debate the evolution of birds. Our understanding of the past is uncertain because the record of the past is incomplete and to some degree inaccurate. Often the evidence that we do have appears contradictory.

Throughout life, people are immersed in uncertainty. They routinely accommodate the uncertainty with a variety of rational, accepting and non-hostile responses. At a simple level, an urbanite might carry an umbrella to meet the possibility of rain; at a more complex level, a farmer might participate in a commodity futures market to protect against the possibility of a drought. Retirement fund managers routinely make investment decisions in the face of considerable long-term economic and political uncertainty, and home and car owners purchase insurance to protect against catastrophe in an unpredictable future. These are all rational actions taken in the face of uncertainty. Nevertheless, there is sometimes a reluctance on the part of decision-makers to take actions addressing complex science-based issues in the face of similar levels of uncertainty, in part because they feel inadequately prepared to contextualize and evaluate the attendant scientific uncertainty. The topic of global climate change illustrates both the scientific complexities and uncertainties, and the difficulties that people and nations have in formulating rational policy addressing the many facets of a changing climate on Earth.

Several themes will run through the chapters of this book, which more or less define my perspectives on accommodating uncertainty, whether ordinary or scientific:

• Uncertainty is always with us and can never be fully eliminated from our lives, either individually or collectively as a society. Our understanding of the past and our anticipation of the future will always be obscured by uncertainty.

- Because uncertainty never disappears, decisions about the future, big and small, must always be made in the absence of certainty. Waiting until uncertainty is eliminated before making decisions is an implicit endorsement of the *status quo*, and often an excuse for maintaining it.
- Predicting the long-term future is a perilous business, and seldom do the predictions fall very close to reality. As the future unfolds, 'mid-course corrections' can be made that take into account new information and new developments.
- Uncertainty, far from being a barrier to progress, is actually a strong stimulus for, and an important ingredient of, creativity.

THE GARDEN OF UNCERTAINTY

Throughout this book, you will be taken on some scientific excursions that will illustrate how uncertainty is woven into the fabric of the scientific enterprise. Many of these treks will be in the Earth and environmental sciences, the field in which I have lived my scientific career. In particular, there will be many forays into that contemporary topic of almost universal interest – global climate change. Probably no other scientific topic has been more regularly in the spotlight during the 1990s than global climate change, and intense debate has swirled around it. The issues of focus at various times have been the reality of climate change, the causes, the consequences, and the political, economic, and social responses to it. As a global scale, complex, slowly developing phenomenon, it displays many of the fascinating facets of scientific uncertainty in general, and it shows how scientists work and thrive in an environment of uncertainty.

The scientific excursions laid out in this book can be thought of as outings in 'the garden of uncertainty', explorations of a vast and irregular tract comprising established plots of annuals and perennials, some newly plowed ground, rare specimens, weeds, thickets, and mazes. Each area of the garden reveals a different facet of uncertainty. And for every insight about uncertainty that one may draw from science, there is usually a parallel and equally revealing experience to be found outside the realm of science that should make readers realize that the scientific world is not so different from their own world. Indeed, science is an important, accessible, and empowering part of everyone's world.

In making comparisons and analogies with the uncertainties that exist in science and in everyday life, my goal is to help readers to understand and accommodate scientific uncertainty in much the same way that they deal with other uncertainties in life. I hope the reader will come away with the feeling that scientific uncertainty should cause no greater hesitation or doubt than do the multitude of other uncertainties that people regularly face and routinely accommodate in their lives. With a better understanding of scientific uncertainty, readers will be able to see through the clouds that sometimes obscure the value and relevance of science to societal issues. In the process of coming to understand uncertainty, they will become more self-confident in grasping what science can and cannot offer.

2 Uncertain about science

This notion that "science" is something that belongs in a separate compartment of its own, apart from everyday life, is one that I should like to challenge. We live in a scientific age; yet we assume that knowledge is the prerogative of only a small number of human beings.... This is not true. The materials of science are the materials of life itself. Science is the reality of living, it is the what, the how, and the why in everything in our experience.

Rachel Carson, in accepting the 1952 National Book Award for *The Sea Around Us*

Science, as Rachel Carson observed, is a part of the very fabric of life. It has its strengths and weaknesses, its successes and failures, its doubts and uncertainties. As scientists attempt to understand how a cell malfunctions to produce cancer, how a gene transmits information to guide an organism's development, how an ecosystem responds to urban sprawl, or how the entire Earth responds to long-term changes in the chemistry of its atmosphere, these investigations are enveloped with uncertainty at every stage. The uncertainty arises in many ways, and the nature of the uncertainty may change through time, but the scientific endeavor is never free of uncertainty.

Has science been debilitated by uncertainty? To the contrary, the successes of science, and indeed there are many, arise from the ways that scientists have learned to make use of uncertainty in their quests for knowledge. Far from being an impediment that stalls science, uncertainty is a stimulus that propels science forward. Science thrives on uncertainty. The uncertainty of how genetic traits were replicated led eventually to discovery of the double helix molecular configuration. Indeed, one might argue that it is *certainty*, rather than uncertainty, that impedes science. The protracted struggle in the seventeenth century by Copernicus, Kepler, and Galileo to overturn the notion that Earth was at the center of the solar system[1] was carried

[1] This history is recounted more fully in Chapter 6.

on in the face of the then-prevalent theological certainty that Earth occupied a very special place in the architecture of the universe.

The uncertainties that scientists face are really not so different from the uncertainties we encounter in everyday life. Risk-taking is extolled in many cultures as an attribute of a successful person. But risk arises precisely because of uncertainty. The willingness and ability to formulate and take action and accept risk in the face of uncertainty is considered a character strength. To be sure, there are risks taken that later prove unwise, but without risk-taking there is an implicit acceptance of the *status quo*. An unwillingness to be motivated by uncertainty is indeed a real barrier to progress.

Ironically, people who are not scientists often equate science with certainty, rather than uncertainty. They have been conditioned by the highly precise and accurate predictions of eclipses, of the daily progression of ocean tides, of the exact times of the local sunrise and sunset, of the clockwork precision of a spacecraft landing on a distant planet. Another aspect of certainty relates to reliability of technology when people pick up the telephone, turn on the television, or turn the ignition key in an automobile, there is an expectation that the device will work. Indeed, when things do not happen as expected or as predicted, there usually is some measure of surprise and discontent. Most people do not relish surprises and are, at some level, uncomfortable with unpredictability and uncertainty.

Certainty in other contexts is a source of contentment. Religious tenets that assure the faithful an afterlife assuage concerns about the abyss of death. Some political mantras, such as 'smaller government is better government' or 'there is no such thing as a good tax', relieve those who recite them from the burden of evaluating a wide range of public policy issues. Recasting a world full of shades of gray into a simpler and starker entity comprising only blacks and whites eliminates the difficult task of weighing nuance and replaces it with the comfort that certainty offers.

When scientists cannot demonstrate a high level of certainty in their understanding of complex natural systems, there is sometimes an undercurrent of impatience and discontent in the general public. In late 2001, bioterrorism in the form of anthrax spores appeared in government buildings and postal facilities in the USA. For a period of time, however, there was uncertainty and confusion in the public health community and at the National Center for Disease Control as to how exactly anthrax might be transmitted, what spore concentrations could be considered hazardous, and how anthrax spores could be rendered impotent. The public wanted answers that public health practitioners could not immediately provide. Similarly in the UK, an outbreak of foot and mouth disease in 2001 was met with a range of scientific opinion as to how it should be contained. Massive culling of neighboring herds was the containment strategy adopted, but scientific opinion was far from unanimous. Long after the disease waned, debate continued about whether the culling strategy was necessary or effective.

When scientists acknowledge that they do not know everything about a complex natural phenomenon such as the spread of disease through an ecosystem, the public sometimes translates that to mean that scientists do not know *anything* about the subject. That, in turn, leads to a loss of public credibility in the capabilities of the scientific community. A byproduct of the loss of credibility is an all-too-frequent willingness of the general public to entertain flimsy pronouncements from kooks, charlatans, and marginal skeptics. With an air of scientific authority and certainty, these pseudo-scientists make assertions that have never been subjected to the rigorous probing that is the foundation of genuine science.

Fortune-tellers, palm readers, clairvoyants, astrologers – the list could go on and on – all thrive on the inability or unwillingness of their clients to recognize the total lack of logical underpinnings and scientific observations in support of these practices. There is absolutely nothing that lends these charlatans any credence whatsoever. But their pronouncements are always carefully crafted to leave their clients with the impression that extraordinary powers have been objectively exercised. In the next chapter, I describe a particularly egregious example of this, a prediction of a major earthquake that was taken far too seriously by far too many people who should have known better.

There are, of course, serious scholars who challenge the notion that science is the only pathway to universal truths. One school of philosophy, loosely referred to as postmodernism, questions whether scientists are neutral and objective, and whether scientific knowledge is truly the outcome of unbiased rational thought. In extreme form, it questions whether a deterministic natural world exists outside of the mental constructs that humans erect. This perspective from the fringe views science as a game with a set of rules created by scientists, and argues that the apparent successes of science in understanding the natural world would not be defensible if we did not accept the rules of the scientific game. A subtheme of this position is that science is a self-serving concept and entity.

In 1996, the postmodern perspective was brought into sharp focus, and ridicule, when Dr. Alan Sokal, a professor of physics at New York University, submitted an article[2] for publication to a journal known to espouse this particular philosophy. The contribution carried the title *Transgressing the Boundaries: Towards a Transformative Hermeneutics of Quantum Gravity*, which seemed to convey a postmodern flavor. Because a physicist had submitted the manuscript, the editors of the journal welcomed the opportunity to publish an article by a scientist that seemed to erode the foundations of science from within. But the article by Sokal was a Trojan horse, a cleverly crafted hoax that illuminated not the philosophical frailty of the scientific method but rather the gullibility of the editors. Sokal had written a seemingly erudite essay, using convoluted language and structure, that really was nothing more than nonsense cloaked in pseudo-scientific jargon. The over-eager editors took the bait and published Sokal's article. Once it was in print, Sokal revealed the hoax. The implications of 'l'affaire Sokal', as it has been dubbed, are many, but for my purposes here the principal point is this: there are people,

[2]Alan D. Sokal, Transgressing the Boundaries: Towards a Transformative Hermeneutics of Quantum Gravity, *Social Text*, 1996.

educated and not, who simply believe that science has nothing special to offer. They are skeptical of, or simply ignore, scientific results.

There is another type of person who may accept scientific results in general, except when the science conflicts with other beliefs they hold dearly. While writing this book, I read the obituary[3] of Charles K. Johnson, president of the International Flat Earth Research Society. Aside from this particular obsession about the shape of the planet, Mr. Johnson seemed to have led a rather normal life as an airplane mechanic. His disagreements with the scientific community were few, except as they related to the shape of the Earth. The image of the spherical Earth taken by the Apollo astronauts from the moon was easily explained: the moon landings were an elaborately staged hoax, and the photograph was but a prop in that scam. We may smile at this quaint explanation, but the pool of uncertainty about science is deepened, little by little, by each and every Charles Johnson who successfully draws attention to his particular astigmatic view of the natural world. In 1994, a poll[4] showed that almost one in ten Americans thought the moon landings were faked. And Hollywood does not help matters with creations such as the 1998 film 'Wag the Dog', in which a US President seeks to divert attention away from personal impropriety by manufacturing a fake war against Albania, including a staged invasion with faked film footage depicting destruction and carnage.

A more widely known conflict between science and personal belief centers on the biblical account of creation in the Book of Genesis. The issue is whether the bible is literally true, word by word. Did God create the entire universe and every living creature in just six days? Geologists and evolutionary biologists make a persuasive case that not all modern life forms were present at the birthday of Earth, and that most of today's life has evolved from other life forms over the vast expanse of geologic time. But biblical literalists do not accept an iota of departure from the Book of Genesis. If Genesis is literally correct, then modern geology and biology must be wrong.

[3] *New York Times*, 25 March 2001.   [4] Marc Fisher, *Washington Post*, 20 July 1994.

Creationists have now taken on the task of proving the tenets of evolutionary biology incorrect, through an endeavor they identify as 'creation science'. The so-called creation scientists have tried to identify flaws in the logic or observations of evolutionary biology so as to 'disprove' it. They have not, however, applied equal vigor to testing the hypothesis set forth in the Book of Genesis. They will not even acknowledge that the account in Genesis is even an hypothesis, let alone testable. They can conceive of no experiment, no observation, that might disprove Genesis. Therein lies the reason that the practitioners of 'creation science' are not really scientists. Creationists will never concede their fundamental position, that all living things are the direct and simultaneous creations of a supreme being. They cannot permit themselves to admit the possibility that the biblical account of creation might not be true or may someday be shown to be untenable. Practitioners of genuine science, by contrast, easily admit uncertainty and are very comfortable working in an uncertain environment. In real science, few concepts can ever be accepted as unquestionably true or absolutely certain.

Indeed, genuine science operates on the assumption that a concept *can* be shown to be false. Falsification occurs when a concept is shown to be logically inconsistent or runs counter to direct observations. Lynton Caldwell, in a review of Michael Zimmerman's book *Science, Non-Science, and Nonsense*[5], describes science as a process of "separating the demonstrably false from the probably true".[6] It is a fundamental underpinning of science that only falsehoods, not truths, can be proven. Truths are simply the survivors of multiple attempts at undercutting. In fact, science progresses in part by continually probing for the soft underbelly of concepts that may have some partial success in explaining some natural phenomena. The unending search for weaknesses may reveal subtle inconsistencies that ultimately require revision or rejection of the original concept.

[5] Johns Hopkins University Press, Baltimore, MD, 1995.
[6] *The Environment*, vol. 38, n. 6, p. 25, 1996.

PEER REVIEW

What is the environment in which these scientific confrontations take place? At the center is a process called peer review. When scientists wish to tell the world about some research that they have conducted, there is an established path to follow. Often the first step is to make an oral presentation of their research at a professional society conference. This requires the prior submission of a very short written summary of their contribution to the committee organizing the program of the conference. This summary is then published in the program so that others may decide whether they want to attend the presentation. At the conference the author will typically make a ten to twenty minute presentation of his or her work, after which there may be questions or discussion from the audience. The opinions expressed in the discussion range widely: agreement, disagreement, skepticism, praise, ridicule.

Should the scientist feel sufficiently encouraged by the discussion in the oral presentation, he or she may then prepare a longer written report of the work and submit it for publication to a scholarly journal. The editor of the journal, in turn, sends the manuscript to other practicing scientists working in the general area of the submitted contribution, asking their opinion about the suitability of the work for publication. The peer reviewers are asked to assess the work from a number of perspectives. Is the work novel and original? Is the methodology employed suitable for the research purpose? Are there errors in the experimental design or in the theoretical derivations? Do the conclusions follow directly from the observations or data presented? What is the level of uncertainty that accompanies the results? This vetting of research reports by experienced practitioners acts as a filter that rejects flawed research but allows research that meets a certain standard to be published for others to read, evaluate, contest, or replicate. Virtually all research articles that are published in professional journals have passed the test of peer review.

The peer review process is not infallible, but the successes of peer review in filtering out weak or flawed science far outnumber the occasional failure. Sometimes peer review will give the benefit of the doubt to a particularly important claim that later proves incorrect, but the process allows for self-correction. In 1999, a team of physicists presented experimental evidence for the existence of a new super-heavy element, number 118 in the periodic table of the elements. Experimentalists in other laboratories, as well those in the original group, tried to reproduce the result by repeating the experiment, but with no success. After two years of failure, the original team published a withdrawal of their claim, acknowledging that they may have misinterpreted the data in their first experiment. Again in early 2002, a paper was published in a very prestigious journal that claimed to observe evidence of nuclear fusion as small bubbles formed and then imploded in an organic solvent when excited by sound waves.[7] In the peer review process, the paper proved to be very controversial, but because the outcome of the experiment, if true, had such extraordinary implications the editors decided to publish the paper. To be sure, the experiment will be repeated in many other laboratories by scientists keen to verify or invalidate the reported results.

A media newcomer, the Internet, has presented a significant challenge to peer review. Anyone with a computer can place his or her research, sound or flawed, relevant or irrelevant, significant or trivial, into the public domain for anyone to read. This places a much greater burden on the consumer of this research to review and evaluate it. The gate-keeping role of peer review that filters out flawed research and prevents it from being published in the scientific journals now falls to every individual reader surfing the Internet. The Internet is a great leveler in that anyone can post almost anything, but the task of deciding whether what is posted has any truth or value falls to the individual user. In earlier times prior to the development of the Internet, the opportunity to make available one's thoughts and ideas to the general public without passing editorial review was a privilege available only to the wealthy, who could self-publish via a vanity press.

[7]Taleyarkhan, R. P. et al., Evidence for nuclear emissions during acoustic cavitation. *Science* vol. 295, pp. 1868–1873, 2002.

The shifting of the burden of evaluation to every individual browsing the World Wide Web makes a public understanding of science and uncertainty ever more imperative.

SOWERS OF UNCERTAINTY

People who do not like what science is telling them often mount subtle and not-so-subtle assaults on science. These take the form of attacks on particular research outcomes that they find threatening. They often argue that had the science been 'properly' conceived and executed, a different result (implicitly meaning one more to their liking) would have emerged. The code-words that frequently identify this particular attack on scientific credibility are 'unsound science', 'unsettled science', 'uncertain science', 'poor science', 'junk science', and the like. What distinguishes these criticisms from those leveled by peer review is that they take place outside of the usual scientific channels and standards. These criticisms appear in newspapers via paid advertisements and letters to editors, and through participants on radio and television talk shows.

These code-word descriptions are used regularly by the petroleum and coal industries as they comment about the causes and consequences of global climate change. In a series of prominently placed op-ed advertisements, the ExxonMobil Corporation[8] frequently denigrates scientific research that documents climate change or that offers evidence that the use of fossil fuels[9] may be contributing to the change. As one of the largest of the international oil companies, ExxonMobil has a strong interest in forestalling a turn away from fossil fuels, and accordingly it has tried to slow legislation or derail international treaties that might limit emissions of carbon dioxide and other 'greenhouse gases' to the atmosphere. One might imagine that if the fossil fuel industry had significant scientific observations that

[8]See, for example, the *New York Times* for 23 March, 10 August, and 21 September 2000.
[9]Fuels such as coal, petroleum, and natural gas are called fossil fuels because they were formed long ago by geological processes. They reside in the rocks making up the crust of the Earth, and they are extracted by mining or pumping from the surface.

contested the role of greenhouse emissions in climate change, they would fight the battle in the scientific coliseum, the peer-reviewed journals where scientific debate routinely occurs, rather than in the media or on the streets. But the fossil fuel industries are more interested in winning the political battles in London, Berlin, Washington, and the state capitals; they spend lavishly in the public arena to confuse and thus undermine public confidence in scientific results.

The strategy of casting doubt and uncertainty about science to influence highly placed decision-makers has not been in vain. In March of 2001, Christine Todd Whitman, the newly appointed Administrator of the Environmental Protection Agency (EPA) in the George W. Bush Administration, abandoned the more stringent limits on arsenic in drinking water that had been promulgated by the previous administration and began a re-evaluation of the scientific basis on which those rules had been framed. "We will use strong science ... to determine what the new limit should be."[10] Such a statement had only one purpose: to undermine public confidence in the previous scientific research on which the newly rescinded regulations had been structured. Those regulations were preceded by more than a decade of reviewing the science addressing arsenic in the environment and its effects on public health, and the publishing of a report on arsenic from the US National Academy of Sciences. That apparently was not sufficient to overcome the opposition from the mining industry, which discharges arsenic as a byproduct of certain types of ore processing, and from communities that would need to upgrade their purification systems if they wanted to continue to drink well water. Of course the EPA only echoes the position of the White House. "We're going to make decisions based upon sound science, not some environmental fad or what may sound good" said President George W. Bush to a group of Environmental Youth Award winners gathered on 24 April, 2001 for a ceremony in the State Dining Room of the White House.[11] Six months later, the National Academy of Sciences, after reviewing the evidence

---

[10] *New York Times*, 21 March 2001.
[11] http://www.whitehouse.gov/news/releases/2001/04/20010424-1.html

again at the request of the Administration, confirmed that new lower limits for arsenic were entirely justified, indeed perhaps not stringent enough. Of course, no new scientific data, no additional 'sound science' had appeared to support the implication that the previously promulgated revisions had been based on unsound science.

These attacks on science are hardly new phenomena. In 1952, when Rachel Carson asserted that the widely used pesticide DDT was having a devastating effect on avian reproduction,[12] the pesticide industry derided her position as being based on weak science. For decades, the tobacco industry denied there was any scientific evidence that showed that smoking was hazardous to health. In the 1970s when the debilitating health effects of lead in the environment came to be recognized, the producers of gasoline that contained lead ridiculed the science. When acid rain in the northeastern states of the USA was found to be a consequence of burning high-sulfur coal in electrical power plants in the Midwest, the electrical generating industry scoffed at the research. When confronted in the 1980s with allegations that chlorofluorocarbons (CFCs) were destroying stratospheric ozone, the chemical industry argued that the science behind the allegation was weak and inconclusive:

> The international chemical industry vigorously denied any connection between the condition of the ozone layer and increasing sales of CFCs. Industry forces quickly mobilized their own research and public relations efforts to cast doubt on the theory.[13]

While the public relations campaign confused the public, the science stood ever firm. In 1995, the Nobel Prize for chemistry was awarded to Sherwood Rowland, Mario Molina, and Paul Crutzen for the research that shed light on the mechanism by which CFCs caused ozone

---

[12] *Silent Spring*, Houghton Mifflin, New York, 1952.

[13] Footnote reference to Dotto and Schiff, *The Ozone War*, pp. 149–165, by Richard Elliot Benedick in *Ozone Diplomacy*, p. 12, Harvard University Press, Cambridge, 1991.

depletion. This was the first and only time the Nobel Prize has recognized research in environmental chemistry.

Does all peer-reviewed scientific research qualify as great science? Of course not. I read scientific journals regularly, submit research reports for publication, and do peer review for them as well. Most scientists will acknowledge that along with the abundant significant research results, the journals contain some correct but trivial contributions, and a few others that later prove to be flawed in methodology. Occasionally, but very rarely, even a fraudulent submission, describing work never done or results never achieved, slips through, only later to be unmasked when someone cares enough to question and check it. To be sure, scientists do not want unsound, careless or poor science cluttering the journals and confusing the state of knowledge. But *uncertain* science, *unsettled* science, is hardly the same as unsound science. The normal state of affairs in science is unsettled and uncertain, and no amount of new research will completely eliminate uncertainty. As earlier questions are answered, new questions appear. Lest this sound like a treadmill of futility, let me assure you that it definitely is not. Far from being frustrated or debilitated by uncertainty, scientists derive strength and creativity from uncertainty. Uncertainty is a challenge, a catalyst for scientific progress.

"PEOPLE LOVE SCIENCE. THEY JUST DON'T UNDERSTAND IT"

Why do so many people have such a hard time accepting and accommodating scientific uncertainty? Are there deeper reasons that go beyond the comforting certainty of religious faith, the apparent certainties offered by charlatans, or the confusing smokescreens floated by some industries trying to protect their economic interests? Much of the problem, I believe, lies in the fact that most people lack an elementary understanding of science generally. This scientific illiteracy provides fertile ground for the appeal of certainty and the confusion of uncertainty to take root.

C. P. Snow, in his famous book *Two Cultures*,[14] outlined the gulf of understanding that separates science from the arts and humanities in the modern university, and in society generally. This view was foreshadowed by Rachel Carson in the quotation that opens this chapter. A 1996 article[15] addressing science education in America began with the statement, "Americans love science, they just don't understand it". Indeed, one can often hear the pessimistic view that the general public will never understand science, let alone the subtleties of uncertainty. Were we not walking on such thin ice of scientific understanding, would we be so vulnerable to the pronouncements of kooks or the smokescreens of confusion laid out by special interests? If we were not so unfamiliar with science, perhaps such obfuscation would not take hold so easily.

The problems with understanding science begin very early, with some inadequacies in the educational system. In a very important sense, children are born as natural scientists. They emerge into a strange world and are curious about everything surrounding them. They look, they touch, they listen, smell, and taste. They make observations of this new world, and they process and evaluate the stream of information coming at them from every direction. They explore, experiment, and learn from their mistakes. Then they go to school.

Schooling in the USA, at least as far as scientific inquiry is concerned, introduces children to a new methodology. The new methodology focuses on science not as a continuation of the curiosity and explorations children make as toddlers. Rather, science in school is, more often than not, presented as a recitation of accomplishment rather than as a process of inquiry. Facts are paramount. Students are told the world is round; Earth orbits the Sun; there are 365 days in the year; insects have six legs; the Amazon is the world's biggest river; Mt. Everest is the highest mountain; rocks can be segregated

---

[14] *Two Cultures and the Scientific Revolution*, Cambridge University Press, 58 pp., 1963.
[15] Michael Carlowicz, *EOS Transactions of the American Geophysical Union*, 27 August 1996.

into igneous, metamorphic and sedimentary categories; atoms have protons, neutrons and electrons. The new emphasis is on stuffing little craniums full of 'facts' that someone has determined every well-educated person must know. Science is presented as answers rather than as questions. Relegated to the distant background is the process of inquiry, of how 'facts' are determined, of how durable or transient 'facts' may be, and of how certain or uncertain we believe them to be.

Answers, as I noted earlier, are to some people more comforting than questions. Uncertainty in a simple context might translate into "It could be this or it could be that", but such a perspective is often seen as being dangerously close to the pit of cultural and moral relativism, where shades of gray between right and wrong can lead young minds astray. "No", say the cultural absolutists, "there are things of which we are certain, and don't try to confuse the issue with uncertainties that only obscure the truth."

TESTING, TESTING ...

The success of schooling is often measured with standardized tests administered to students locally and nationally, to gauge achievement in reading, math, and science. Some school systems and their teachers are ranked according to the success or failure of students in such standardized tests. Proposals are regularly floated to link the governmental funding of schools to their performance on standardized tests. It should come as no surprise that some schools now 'teach to the test', recognizing that their political and perhaps financial support may depend on doing well on these tests. From an international perspective, however, even teaching to the test has not produced dramatic results from US schools. The eighth graders (age thirteen students) in 1999 scored below the international median in both science and mathematics in the Third International Mathematics and Science Study (TIMSS), which tested students in twenty-three countries.[16] Within the USA alone, the National Assessment of Educational Progress

---

[16]http://ustimss.msu.edu/

every few years administers a mathematics test to fourth, eighth, and twelfth graders (ages nine, thirteen, and seventeen, respectively), with results categorized as 'below basic', 'basic', 'proficient', and 'advanced'. In the test given in 2000, only one in three nine- and thirteen-year-old students and fewer than one in five seventeen-year-old students reached the proficient level.[17]

I recognize of course that scientific progress, indeed progress in any of life's endeavors, must have an educational foundation that includes basic literacy and numeracy. Reading, writing, and quantitative skills surely must be included in a list of life's essentials. But as necessary as they are, if they alone are the targets of education, we will shortchange both the students and the society they are a part of. Other important skills – how to observe carefully, how to think critically, how to deal with conflict, how to develop teamwork – are not easily tested but arguably are equally important, or more so, to the success of students and to their community.

This emphasis on acquiring 'knowledge' persists throughout the primary and secondary educational systems and continues unabated in many higher education curricula. Textbooks for the introductory survey courses in science too often are dull compendiums of what we *do* know but without a stimulating summary of what we *don't* know. Where are the frontiers of science described in these textbooks? What are the unanswered questions that might excite imaginative students and rekindle the natural curiosity they had as young children? Why do they not learn of the uncertainties in the field?

The history of how a field of science has evolved over time can be revealing of the false starts, the blind alleys that scientists followed in times past. Antonio Machado, a Spanish poet of the early twentieth century, captured this idea when he wrote: "Traveler, there is no road. You make the road as you go." Although the history of a discipline is usually not couched in terms of the uncertainty that enveloped the field, it can highlight the conventional wisdom of a

---

[17] *New York Times*, 3 August 2001, p. A21; *New York Times*, 21 November 2001, p. A12.

certain time and show how, in the face of conflicting observations and competing ideas, that conventional wisdom began to unravel, only to be replaced with newer concepts. What the perspective of time and history offers is an opportunity to see how science as a field of *inquiry* has evolved, and how probing questions and critical thinking contributed to better understanding. Without any historical context, students must settle for a snapshot of today's answers, not yesterday's or tomorrow's questions.

In the graduate degree programs, where in principle we train the professional scientists and future professors, universities must try to undo all of this. By the time students reach graduate school, they have focused far too long on giving answers instead of asking questions. They have a hard time formulating a research project that poses an interesting non-trivial question, and that lays out a pathway that may shed some light on it. Many of my scientific colleagues in the university are not helpful in freeing students from the educational constraints that have dulled their curiosity. Although many faculty members themselves have a vision for their research, too often they view graduate students as cogs in their personal research machine. The student is not asked to formulate a research question and an approach to answering it. Instead they are assigned a project, narrow in scope, seldom explained in the context of the larger research vision. The students are instructed on how to make use of the most sophisticated research equipment to measure this or that but are left out in left field as far as the relevance of the measurements is concerned. They learn a lot about how, but little about why. And, of course, when the technical skills they have acquired are made obsolete with the next generation of instrumentation, many will fall away from science disillusioned.

––––––––––

So why do we have so much science illiteracy? Why are people so susceptible to simplistic ideas and false assertions? Why are they puzzled by scientific uncertainty? In part, I think it is because the science education that most students receive stifles their natural scientific instincts. Many students lose interest in science in the primary and secondary schools because it does not take advantage of their natural curiosity. The higher educational system then perpetuates the problem, graduating science 'majors' who have absorbed all the 'facts' but who are not equipped to challenge them. The same stultifying system trains new elementary and secondary school teachers, who repeat these patterns, and at the postgraduate level trains new scientists as capable executors but not imaginative formulators. In short, educational practices common in many countries have led to a widespread adult population that is interested in and yet puzzled by science, principally because they do not understand how scientists go about the business of asking questions and evaluating answers.

Continuing the metaphor of the garden of uncertainty, we have started our tour in the orientation pavilion, where the displays have illuminated some of the sociological, political, and educational facets of science and uncertainty. The next chapter remains set in the orientation pavilion, where we will focus on a special institution that stands between science and the public: the mass media. Can the media help to convey science to the public, not simply in terms of accomplishment, achievement, and certainty, but rather as a process or method of inquiry that is stimulated by failure and which flourishes in the dim gray light of shadows cast by uncertainty?

3 Can the media help?

Science is a long movie, and the news media generally take snapshots.

John Schwartz[1]

Is it really essential that the public understand science? Why not let scientists do their thing, and let the rest of the world get on with their business too? Unfortunately, in the modern world, that is a path we can ill afford to follow. Whether we realize it or not, science is too much a part of the fabric of our lives to be shunted aside as a curious sideshow. The economy, national defense, environment, and our health are more than ever before dependent on scientific progress. The emergent role of information technology in our economic productivity, the feasibility of a ballistic missile defense shield, the human contribution to climate change through the combustion of fossil fuels, the implications of the newly mapped human genome all should be reminders that we cannot divorce ourselves from science, even if we might like to. And yet for all of the obvious relevance of science to our daily lives, many people remain ill equipped to assimilate much beyond the rudiments of science.

If, as I have argued in the previous chapter, our schools have generally failed to develop an awareness and appreciation of science, one can envision a second line of defense against scientific illiteracy: scientists working closely with the mass media to inform and educate the public. When issues of scientific understanding or misunderstanding arise, should we not be able to turn to television, radio, newspapers, magazines and the Internet for clarification and insight? With their billion dollar budgets, talented staffs and sometimes close working relations with practicing scientists, the potential for making science accessible would seem high. Both scientists and journalists are generally well educated and have similar intellectual foundations:

---

[1] *Washington Post*, 21 February 1999.

inquisitiveness, skepticism, and an ability to piece together a story from incomplete and sometimes inaccurate information. Surely, the science education that has been left undone by the schools can later be remedied by scientists and the media.

That is a big responsibility for both scientists and the media, and unfortunately one for which they are both generally unprepared.[2] Scientists are frequently uncommunicative, the media are impatient and internally competitive, both groups misunderstand and to some extent mistrust each other, and neither typically feels a strong responsibility to educate the public about science. Add to that mix the fact that there are many forces of obfuscation at work, for example, the special-interest groups such as the tobacco and fossil fuel industries, that do not want to have certain scientific issues clarified, nor uncertainties placed in context and evaluated. But within the media, there are also forces at work that effectively undermine scientific understanding. Many talk-show hosts on radio and television consider their first role as one of entertainment; on the rare occasion when some science makes its way onto the chart, it usually fares poorly in the give and take of talk. It is not that the hosts at the outset intend to make the science unclear, but many are unwilling to invest the time to understand the complexities. They definitely do not want science, nor any other topic they feature, to be complicated or burdened with shades of gray. Simplicity laced with dismissiveness is a good formula for entertaining banter and, in the hands of a glib talk-show host, science is often quickly reduced to rubble.

Other weaknesses abound in this envisioned media–science educational alliance. The problems of coalescing these potential collaborators in common cause are multifaceted, and failures by both groups strew the field. As the old saying has it, there is enough blame to go around. Let us first see what obstacles the scientists erect.

---

[2]A thorough discussion of the relationship between scientists and media professionals is presented by Jim Hartz and Rick Chappell, in *Worlds Apart: How the Distance between Science and Journalism Threatens America's Future*, First Amendment Center, Vanderbilt University, 1998.

THE SCIENTISTS

Journalists need sources. They cannot report about science if scientists will not talk to them. And scientists like their work to be recognized. Can they get significant recognition other than from the media? Unfortunately, at least from the perspective of science education, the answer to this question is clearly 'yes'.

The principal form of recognition that scientists seek is recognition *from their peers*. This comes in the form of publication of their research results in peer-reviewed scientific journals. It comes from research grants awarded on the basis of peer-reviewed competitive proposals. It comes from salary increases and promotions bestowed on the basis of peer evaluations. And for a very few, it comes from winning prestigious awards such as the Fields Medal in mathematics, the Crafoord Prize in earth science, or the Nobel Prize in physics, chemistry, medicine or economics. But in general, the reward system for academic scientists does not place much value on engagement with the non-academic world. In fact, there is an underlying feeling that in an academic career, advancement is retarded by spending time in non-academic endeavors. Carl Sagan, the Cornell University astronomer and prolific author who brought so much science into the popular realm, was never elected to the National Academy of Sciences, one of the highest forms of peer recognition a scientist can achieve in the USA, equivalent to becoming a Fellow of the Royal Society in the UK. Speculation about the reasons for denying him this recognition have centered on his extraordinary success as a popularizer and expositor of science, both as an author and on television. Many accomplished scientists place a low value on, if not outright disdain for, such media endeavors.

Although clearly in the minor leagues when compared with Carl Sagan or Stephen Jay Gould, I do attempt to bridge the gap. In addition to substantial teaching and engagement in research, I have made considerable effort to write about science for non-scientific readers, to speak about science to non-scientific audiences, and to be available to the media to help them to convey to the public the significance of my research (or that of others) when it is published. I have received nice notes from the Academic Vice-President of my university, thanking me for taking the time to work with the media (of course my university's name also appears in the articles or on the air). I recall a conversation with Dr. Neal Lane, Director of President Bill Clinton's Office of Science and Technology Policy and former director of the US National Science Foundation, in which we discussed my concept for this book. He urged me to get to writing straightaway, commenting that helping the public to understand science better was an urgent matter. I have received a personal note from the Vice-President of the United States, thanking me for presenting a seminar to congressional staffers on aspects of global climate change.

To be sure, there is a lot written and said about the importance of helping the public to understand complex scientific issues. However, scientists working in universities, where most of the scientific research is undertaken, are subtly discouraged from reaching out to the media by the nature of the reward system. When it gets down to a merit evaluation by colleagues as part of the annual salary-setting considerations, public engagement counts very little. The well-worn joke that all of the non-academic recognition plus a couple of dollars will buy you a good cup of coffee is hardly irrelevant. The issues that count (and I use the term literally) are how many research papers one publishes, how many graduate students and postdoctoral scholars one supervises, how many undergraduates one teaches, and how many research grants one garners. Broader issues of general education, such as those addressed in part in this book, and the more public outreach activities that form a part of my professional life, seldom count for much in university merit evaluations. So it should come as no surprise that, if scientists receive little tangible encouragement to work with the media to make science accessible to a broader non-academic audience, they seldom make the effort.

Even if a suitable reward system were in place, training scientists to be media-friendly is hardly a straightforward task. Perhaps at the head of the list of difficulties is the fact that scientists and journalists do not ordinarily work in the same time frame. Science does not typically generate daily events such as a mine collapse, a cricket match, a political debate, or a ballet performance to capture the attention of the media, and scientists seldom face daily deadlines.

Perhaps the closest thing to making an extended scientific endeavor into a news event is the occasion when research results are presented at conferences or published in professional journals. We are now becoming accustomed to news reports that begin "In a study published today in the *New England Journal of Medicine*...", or, "In a presentation this week at the meeting of the European Geophysical Society...". But how do media professionals recognize truly important contributions among the many thousands of articles published and papers presented each year? You can be certain that very few reporters have a regular assignment to read the *Lancet*, or the *Journal of the American Medical Association* or the *Journal of Geophysical Research*, a task that for most journalists would be roughly equivalent to death by slow torture. However, in recent years, the universities, the professional societies, and the publishers of the scientific journals have mounted an impressive effort to build bridges between the scientists they know and love, and the media professionals who will convey the science to the public. This is a significant departure from the isolationist traditions of most academic scientists and their professional societies.

My own university, mindful of tending to its public image, has an Office of News and Information Services with a very talented staff. They pro-actively ask science department chairs for a calendar of the professional conferences that faculty attend, and they sit down with the department chairs to identify the presentations that have particular significance. An interview with faculty making the presentations may follow, and a press release is prepared for distribution to the media. Sometimes this leads to interviews with reporters following the actual presentation.

The scientific journals themselves also have an impressive press machine at work, to call to the attention of the media the important advances appearing in the pages of their publications. Two of the most prestigious international scientific journals are *Nature*, published in the UK, and *Science*, published in the USA by the American Association for the Advancement of Science. These journals, read widely in the scientific community, appear weekly and feature fifteen to twenty research reports authored by scientists from around the world. Prior to publication each week, both *Science* and *Nature* provide press releases about the various articles, editorial commentary about a selected few, and contact information to enable the press to reach the scientists involved in the research for additional perspectives. After colleagues and I recently published a research report in *Nature* on how global warming was evidenced in rock temperatures,[3] newspaper, television and radio interviews consumed almost a week with hardly a break.

However, all of the efforts by institutions, professional societies, and publishing houses cannot bring science to the public without the cooperation and engagement of the scientists themselves. The ultimate responsibility for removing the curtain of obscurity that surrounds science lies with the practitioners. They must be available and be effective communicators with the print and electronic journalists who want to help them to share with the public what goes on in the house of science. For those scientists working in universities, the latter should not be an insurmountable barrier. They face the task of conveying science to classrooms full of students all the time. Successful teachers do not hide behind a barrage of technical jargon. They have learned to organize and simplify their material, to help students to see the forest as well as the trees. Many have recognized that it is more important to convey science as a process of inquiry rather than a catalog of achievement. They know that all science is tentative and uncertain, and that the uncertainties spur creativity and drive science forward. Working with the media requires the same principles as engaging students in the classroom. Scientists must be alert and

---

[3] Huang, S., Pollack, H. N., and Shen, P.-Y., Temperature trends over the past five centuries reconstructed from borehole temperatures. *Nature* vol. 403, pp. 756–758, 2000.

sensitive to possibilities of misinterpretation, and they must make efforts to say clearly what their work means, as well as what it does not mean.

Journalists need scientists not only for material but also for insight. It is one thing for a reporter to see in a journal a scientific report with some arcane title, and quite another thing to recognize the significance (or insignificance as the case may be) of the report. While surely there are exceptions, generally it takes someone who is an active researcher to be able to evaluate research. That is, in fact, the basis of the peer-review system, which evaluates and filters scientific contributions before they see the light of the printed journal page. This same system of review also guides the funding of research by federal and state agencies, by seeking the advice of active scientists about the strengths and weaknesses of proposals that have been submitted seeking funds to conduct scientific research. I also believe that teaching at the university level is enhanced by active engagement in research. No one is better equipped than a researcher to recognize the robustness of a certain body of experimental data, or to identify the soft underbelly of a theory. No matter how voracious a reader and synthesizer of the scientific literature one may be, there is nothing quite equivalent to having been in the scientific line of fire.

Journalists are busy folks facing deadlines, and they appreciate a scientist who can cut to the quick. They are neither eager to hear nor skilled at sorting through an endless array of qualifications that scientists may weave as a protective cocoon around their results. And while journalists may be aware of the fact that there is some uncertainty about the conclusions, they also will appreciate the scientist who can place the level of uncertainty in some familiar context. Roberta Hotinski, a graduate student in Earth Science at Pennsylvania State University, spent the summer of 1999 in the newsroom of *US News and World Report* as a Mass Media Fellow in Science and Engineering, sponsored by the American Geophysical Union. In writing about her experience she offered the following advice. "If you can express your level of certainty in terms of odds or common analogies, reporters will have something concrete to emphasize. For example, you could characterize your certainty that global warming is upon us as comparable to your belief that a) the sun will rise tomorrow, b) your kids will go to college or c) you'll win the lottery."[4] Scientists regularly characterize the uncertainty of their results by defining a quantitative range, called the 'error bar', in which their results sit. Translating error bars into ordinary language that journalists can understand would go a long way toward making research results more accessible.

THE MEDIA

Not all the barriers to a close working relationship between science and the media arise from the scientific camp. Let us note just a few of the impediments erected by the media. As John Schwartz observed in the quotation opening this chapter, the media, with snapshot camera in hand, generally do not have the time or patience for a long movie. Even as a long movie is playing, the media are distracted by the more immediate events, and they may not even realize they are in the theater. Moreover, as topics such as global climate change unfold over decades, there is a tendency by the media to consider it as lacking freshness or currency. That climate change comes to the fore time and time again stems from the fact that dramatic climate-related effects continue to occur and collectively call our attention to the slow changes that are taking place. Another debate in Parliament or Congress is passé, but the sudden separation of an iceberg the size of Scotland from an Antarctic ice shelf is news. Another press release from an international oil company calling for 'more research' on climate change draws yawns, but an ice-breaker discovering that the path through the Arctic Ocean to the North Pole is ice-free pulls the media back again to another facet of climate change. And when a survey of glacial ice on Kilimanjaro indicates that Ernest Hemingway's immortal snows may disappear over the next fifteen years, someone in the newsroom takes note.

[4]Roberta Hotinski, *EOS Transactions of the American Geophysical Union*, 16 November 1999.

Enough snapshots strung together can begin to look like a movie to the public. Eventually, through repetition, these large-scale environmental concepts can become embedded in the public awareness. In 1987 when the Montreal Protocol, an international agreement to phase out the manufacture and use of ozone-destroying chemicals, was taking shape, it was very significant that there had been a decade or more of 'snap-shot' visibility preceding the international conference. Richard Elliot Benedick wrote

... the power of knowledge and of public opinion was a formidable factor in the achievement at Montreal. A well-informed public was the prerequisite to mobilizing the political will of governments and weakening industry's resolve to defend the chemicals. The findings of scientists had to be made accessible and disseminated. ... The media, particularly press and television, played a vital role in bringing the issue before the public and thereby stimulating political interest.[5]

As I have already mentioned, the inadequacies of many primary and secondary schools, and indeed some institutions of higher learning, in the teaching of science and general numeracy have now affected several generations of students. Included are most of our practicing journalists. The scientific education of most media reporters has been little different from their elementary, secondary, and university classmates, and this underlies their subsequent difficulties in conveying the meaning and significance of science to the public. And when scientists themselves offer anything short of a unanimous interpretation, when uncertainty is expressed, media reporters are generally ill-equipped to evaluate the disparate perspectives and to help the public to appreciate and accommodate the uncertainty.

Other factors, not unrelated to the educational shortfall, include an undervaluing of science as uninteresting and seldom newsworthy. This attitude is manifest through a lack of commitment

[5] *Ozone Diplomacy*, Harvard University Press, Cambridge, 1991, p. 5.

of staff and time to science coverage and analysis by publishers, producers, and editors. This attitude is underpinned by the opinion that science is not really competitive for the limited time and space available on a daily basis. In addition, sadly, some media 'gatekeepers' feel that public education is not the responsibility of the media; they believe their assignment is to report and chronicle the events of the day, rather than to interpret and place those events in context. Or worse yet, they simply wish to entertain their readers.

For a profession proud of its skilled use of language, the media are frequently careless in evaluating the language used by special interests, language that colors the way in which issues are portrayed. All too often code-words that subtly distort an issue are unwittingly incorporated into articles. Describing atmospheric carbon dioxide and methane as 'so-called greenhouse gases' creates an impression that perhaps they are not, an impression that absolutely no atmospheric scientist would support. When journalists use a phrase such as 'the greenhouse theory' they create an impression that the atmospheric greenhouse effect is perhaps only a concept and not grounded in reality. Nothing could be further from the truth; Earth's surface has been warmed by the greenhouse effect throughout most of its history. The proper debate is not about whether our planet has a greenhouse effect, but rather about how much the greenhouse effect is changing because of human activity. It is a disservice to the public to repeat the words 'sound science', 'junk science' and 'creation science' without paying careful attention to who is using such language and what their motivations might be.

The media often view controversy as more interesting than the science itself. Scientific debate is viewed almost like a sports contest, with competition sure to yield a winner. But unlike sporting events, to which a very significant fraction of each day's reporting is devoted, science issues do not have the benefit of a large media staff and in-depth analysis. A large newspaper may have a team of sports writers, with specialists in golf, tennis, cricket, football, basketball, and baseball. We may learn in great detail over many weeks about the training regimen, coaching strategy, and group psychology of World Cup football teams, or about deep inner thoughts of a Tour de France bicyclist. But a scientific debate is often reported as a 'he-said, she-said' encounter with little insight provided to a reader or viewer to help them to understand the subtleties. Because journalists wish to avoid being branded as one-sided advocates, and because they recognize their own inadequacies to analyze a situation, they often feel compelled to give equal time to opposing points of view, irrespective of the strength of the scientific arguments supporting one side or the other. The result is often to give unwarranted attention and thereby bestow credence to frivolous pronouncements and marginal debate.

Media coverage going astray
  Before: "Fault Line's Threat Hits Fever Pitch"[6]
  After: "Media at Fault Over New Madrid Quake Scare"[7]

An instructive example of media coverage gone astray was the treatment of the prediction that a major earthquake would strike the mid-continent of the USA, or central California, or Tokyo, or somewhere else on 3 December 1990. This prediction was issued by Dr. Iben Browning, a business consultant whose advanced degrees were in biology, not geology or seismology. In the USA, particularly in the central states, the media jumped upon the prediction and, without making any significant effort to assess its validity or likelihood of occurrence, manufactured a major event, creating a frenzy of local concern that led to school and business closures, evacuations, emergency preparedness drills, insurance scams, and unusual entrepreneurial activity that took advantage of the almost carnival-like atmosphere at ground-zero.

Browning's earthquake prediction was based on the idea that on 3 December 1990 the alignments of the Sun, Moon and Earth would be such that the gravitational pull of the Sun and Moon would reach an

[6]*USA Today*, 28 November 1990. [7]*St. Louis Post-Dispatch*, 8 December 1990.

unusually high level. These extra tugs would add to tectonic stresses accumulating in a seismic zone on Earth to trigger a large earthquake. It is a kind of 'straw that breaks the camel's back' concept. While it is true that the planetary alignment to which Browning called attention did occur, its prediction required no more special scientific skill than the ability to read an almanac. What evidence did Browning have that the locations predicted for the earthquake were loaded to the breaking point, waiting for the last straw? Absolutely none. Apparently he simply selected areas that historically had featured large earthquakes and declared that these regions were ready for another jolt.

Fortunately, 3 December came and went uneventfully, without so much as a noticeable tremor at the anticipated epicenter in southeast Missouri (nor a major event anywhere in the world for that matter). The circus tent at the epicenter folded up, and the media turned their attention to other more 'newsworthy' events in the days following. Could the public have been better served? Absolutely. Journalists had many opportunities to probe more deeply into the scientific foundations of the issue and exercise judgment in their coverage, but generally failed to do so.

What might the media have done? Browning's 'credibility' in earthquake prediction apparently stemmed from a talk he gave to manufacturing executives in San Francisco on 10 October 1989, in which he stated that a major seismological event would occur *somewhere* in the world within a week's time. Sure enough, within the week, right there in California, occurred the Loma Prieta earthquake, leading to sixty-seven fatalities and causing very significant damage in San Francisco. But was this 'prediction' really such an achievement? With only a little investigation, reporters could have easily learned that in an average year there are about 120 large earthquakes (magnitude 6 or greater) around the world, or about one every three days (assuming for simplicity that the quakes would be distributed uniformly in time). Thus Dr. Browning was on pretty safe ground to make a reasonable guess about timing. A cub reporter could be equally prescient.

In addition to specifying the 'when' of an earthquake, seismologists require that a valid prediction must also include the place, the magnitude, and an estimate of the probability that a quake of that size might occur at that location anyway. As far as Browning's designation of location goes, somewhere in the world is surely a safe bet, and forecasting a 'major' event leaves ample wiggle room for interpreting the size of the predicted event. If the target is big enough, there is a fairly good chance of hitting it. As to the requirement of a probability estimate, this can be best illustrated with an example: a 'prediction' that the San Francisco Bay area will experience a magnitude 2 earthquake on a certain day is not considered a prediction of any significance even if the event occurs as predicted. Why not? Because there are many hundreds of magnitude 2 earthquakes in that area each year, and therefore the probability that one will occur on *any* given day is very high. Had Browning in his talk to the business executives predicted a magnitude 7 earthquake for the central California region within the coming week, as opposed to a 'major' earthquake somewhere in the world, then the occurrence of the Loma Prieta earthquake would have dramatically bolstered his stature as a seer. A direct hit on a small target is more impressive than a dart placed somewhere on a big wall.

Why did the media discount the fact that the scientific community was of the virtually unanimous opinion that the Browning prediction had no scientific merit? The US Geological Survey's National Earthquake Prediction Evaluation Council declared that Browning's prediction had no validity, as did most professional seismologists from nearby universities. There was, however, one exception, a seismologist from Southeast Missouri State University, who lent some considerable public support to Browning's prediction. To the media, this lone supporter apparently provided a full counterweight to the massed scientific opinion sitting firmly on the other end of the seesaw. Had the media looked into the background of this supportive seismologist, they would have discovered that he was no stranger to the earthquake prediction business himself; he had collaborated with a psychic in predicting an earthquake for North Carolina in 1974. And had the media delved further into Dr. Browning's own background, they would have discovered that he also attributed the rise of the Nazis to tidal forces! Whatever happened to the journalistic practice of checking the background and credibility of sources?

The media might also have discovered that the science behind Browning's prediction of the 3 December 1990 event, a planetary alignment that exerted incrementally stronger gravitational stress on the region, was also old and unsuccessful science. Because the positions of the Sun, Moon and planets can be so reliably predicted, they have always been an attractive component of other prediction schemes. In natural systems where the physics of a process is well understood, such as the rise and fall of the ocean tides along shorelines and in harbors, the predictions are remarkably successful. But in the business of catastrophe prediction, in which the natural system is complex and poorly understood, the record shows no successes.

The authors of a book[8] that received wide publicity when published in 1974 argued that a special alignment of the planets that would occur in 1982, a configuration that occurs only once every 179 years, would trigger a devastating earthquake in southern California. The physics of the triggering was more sophisticated than that implied by Browning in his 1990 prediction, but the hype was the same:

> Geophysicists report that [the San Andreas Fault] is overdue... and just needs a trigger. There can be little doubt, we feel, that the planetary and solar influence... following the rare planetary alignment, will provide that trigger. In particular, the Los Angeles region will, we believe, be subjected to the most massive earthquake experienced by a major center of population during this century.[9]

Needless to say, the San Andreas fault in southern California did not lurch significantly in 1982, and Los Angeles remains standing. But this

---

[8]John Gribbin and Stephen Plagemann, *The Jupiter Effect: The Planets as Triggers of Devastating Earthquakes*, Walker and Company, New York, 136 pp., 1974.
[9]From the preface of *The Jupiter Effect*.

episode had been largely (entirely?) forgotten by the media by the time Browning gave us his prediction for 1990. The manila folder in the newsroom files labeled 'Earthquake Prediction', available for reporters to consult for background and historical perspective, apparently was empty. I am not aware of a single reference to the Jupiter 'non-effect' at the time of the Browning media frenzy. And when yet another book[10] with a similar theme predicted that global catastrophe would strike on 5 May 2000, the media coverage,[11] while restrained and free of hype, again showed no apparent awareness of how frequently this particular disaster theme reappears.

Why did the media not unmask the Browning prediction as one bordering on the ridiculous? Why instead did they paint him as an unorthodox and unappreciated genius, who had hit upon a marvelously simple strategy for predicting earthquakes? Surely the answers to these questions are not simple, but one aspect may be a deep-seated distrust of expertise and conventional wisdom. In various contexts, this attitude could be described as 'anti-elitism', 'skepticism', 'contrarianism', or as 'rooting for the underdog'. If all the bigwigs say something is not going to happen, won't it be fun to see them eat crow when it does happen? Won't we have a good laugh when someone not anointed by membership in the in-group succeeds despite the opposition of the pooh-bahs? Won't it be nice to have some unrecognized outsider show that something is really very simple, when all the experts have been telling us that it is very complex? Complexity is a real barrier to understanding and can turn people to seek simpler answers, no matter how inadequate or demonstrably false those answers may be.

This attitude was displayed in an interesting way in yet another example of earthquake prediction. In the early 1970s, a scientist working in the US Bureau of Mines (USBM) developed a theory of earthquake prediction based on his studies of how rocks break in laboratory stress tests. He then applied his theory to the real (outside the laboratory) world, and in 1976 predicted that the largest earthquake of the

---

[10]Richard Noone, 5/5/2000: Ice, The Ultimate Disaster, Harmony Books, 1986, 1997.
[11]New York Times, 7 May 2000, p. 26.

twentieth century would take place off the coast of Peru on 28 June 1981.[12] Quite naturally, Peruvians were concerned and sought some evaluation of the merits of this prediction of coming catastrophe. Because the prediction was issued by a US government scientist, Peru initiated discussions with the US through diplomatic pathways. The US government in turn asked the US Geological Survey (USGS), the federal agency charged with assessing seismological hazards, to evaluate the prediction theory and methodology. The USGS evaluation team concluded that there was no validity to the USBM scientist's approach, and no credibility should be given to his predictions. In particular they concluded that the seismic hazard in Peru, already a country visited historically by severe earthquakes, was in no way heightened on the basis of the USBM scientist's concept.

But that was not the end of the matter. Some wondered privately and later publicly whether the USGS rejected the USBM's approach out of jealousy. It was well known that the USGS had invested considerable effort in earthquake prediction research with little success. Wouldn't it be an embarrassment if a single researcher in the USBM found the secret to earthquake prediction when the entire USGS team had failed to do so? Could this be a case of the little guy succeeding in unraveling a complex process essentially in his spare time, when the supposed professionals were still following one blind avenue after another? An unappreciated genius making the experts look incapable? The episode had all the hallmarks of the later Browning saga, including the unambiguous failure of the predicted event to take place. The day came and went without so much as a seismic ripple being felt in Peru. The USBM prediction turned out to be empty, just as the professional seismologists had predicted.

IT'S TOUGH TO GET IT RIGHT

Just as it is difficult to make scientists into media-friendly colleagues, it is equally difficult to make journalists into sophisticated observers of science. We would hope that journalists could be more than just

---

[12]For a full analysis of this episode, see Richard Stuart Olson, *The Politics of Earthquake Prediction*, Princeton University Press, Princeton, NJ, 1989, 187 pp.

literal reporters, that they would be able to offer some insight and perspective. At their best, journalists must be skeptical and ferret out weaknesses and contradictions. Those who do reach a level of understanding and familiarity with science typically comment on what a hard slog it is to reach that plateau of comfort. Malcolm Browne of the *New York Times* remarked, "A science writer must be a perpetual student.... It takes a prepared mind to appreciate the value of a subtle experiment."[13] Walter Cronkite, who covered the US space program for a national television network, overcame his scientific unease through "many long hours of study".[14] But not many journalists are afforded the opportunity to develop background and perspective. More typical is the experience of a new young science reporter, who felt immense pressure when she began her work. Within a few weeks she had to cover stories on pain centers in the brain, the effects of low-frequency electromagnetic radiation on human health, novel techniques for dating sedimentary rocks, a potential vaccine for Alzheimer's disease, fiber optics cable deployment, and forecasting of global warming.[15]

Happily, there are a few notable exceptions: *The Times* of London and the *Guardian* in the UK, the *Globe and Mail* in Canada, the *New York Times* in the USA all offer science coverage well beyond the norm. The completion of the map of the human genome in 2001 was, of course, reported as a lead story nearly everywhere, but the *New York Times* devoted another ten full pages to discussing the medical and ethical implications of this remarkable scientific achievement. In an interval of about a month, the *New York Times* had several different science stories prominently on the front page: the discovery of crisp erosional features on Mars, suggesting geologically recent water seeping from the subsurface; the re-interpretation of some feathery fossils that suggested a different evolutionary origin for birds; the discovery of an elusive particle in the neutrino family; and the sociological

[13] *New York Times*, 27 February 2000.

[14] Jim Hartz and Rick Chappell, *Worlds Apart: How the Distance between Science and Journalism Threatens America's Future*, First Amendment Center, Vanderbilt University, 1998.

[15] Roberta Hotinski, *EOS Transactions of the American Geophysical Union*, 16 November 1999.

consequences of genetic testing. In addition, each Tuesday, the *New York Times* offers an entire section, the *Science Times*, produced by a staff of fifteen science writers.

But is this reason to celebrate? Only a small fraction of newspaper readers in the UK read *The Times* of London or the *Guardian*, and probably a smaller fraction of Americans read the *New York Times*. And each of these newspapers devotes far more space to financial news, entertainment and the arts, and sports on a daily basis than it does to science on a weekly basis. In the *New York Times* on 21 July 2000, there were five important science and science-related articles, a good day for science visibility. However, on that same day there were *thirty-eight pages* devoted to arts and entertainment, and *six pages* to sports. And we must remember that most media outlets do not even employ a science reporter; they simply take stories off the wire services and cut them somewhat indiscriminately to fit the space or time available. Most local television channels do have a 'weather person', someone perhaps with a degree in meteorology. Whenever science intrudes into the daily routine, it often falls to the weatherman or woman to deal with it.

With a touch of sadness, I conclude that the defects in the structure of science education will not be miraculously annealed by enlisting the aid of the media. Neither scientists nor journalists, each for their own reasons, place sufficient value on the endeavor to make it happen. Counting on practicing scientists and the media to take up where science education has left off is, I fear, too grand a dream. The potential may be there, but the realization is, just like a mirage, well beyond the horizon. I am sanguine that both science education and media attention to science will improve, but the turnaround will be very slow.

It is now time to leave the orientation pavilion and step out into the garden of scientific uncertainty. As mentioned earlier, the garden has many domains and plots that reveal the multifaceted character of uncertainty. However, collectively, the many floral displays and untended natural areas, the thickets and subtle mazes, create a mosaic with images that stand out amidst the individual tracts of uncertainty. One by one, each tract will illuminate an aspect of uncertainty; together they will lay out how uncertainty arises, how it is a stimulus for creativity, and why it is akin to a glass half full, not half empty. The next chapter explores how our intuition is often an inadequate guide to understanding complex phenomena, thereby allowing uncertainty to cloak the unfamiliar terrain that exists outside of our everyday experience.

4 Unfamiliarity breeds uncertainty

Experience is a hard teacher because she gives the test first, the lesson afterward.

Vernon Sanders Law

When we experience things in the course of our lives, we become familiar with them, perhaps understand them, and come to accept them as a normal part of life. But when we first encounter something that we have not previously experienced, something we are unfamiliar with, there is a natural tendency toward caution. And if we are presented with an abstraction, something totally outside of our experience, skepticism or even disbelief is not an unnatural reaction.

In this context, uncerainty goes hand in hand with unfamiliarity. What we are unfamiliar with, we are uncertain of. And much of science is unfamiliar ground for many people. Although Albert Einstein thought otherwise, science is really not just 'common sense'. If it were, no one with a modicum of common sense would be puzzled or baffled by it. Science requires a certain amount of abstraction, and the placing of observations into a context or framework. When that framework is one's immediate environment, familiarity and understanding come readily. But when the spatial framework is much smaller, as in particle physics, or much larger as in astronomy, it takes a willing mind to explore this unfamiliar, uncertain terrain. Similarly, there are processes that operate at time scales vastly different from those of everyday human experience. The apparently instantaneous completion of a chemical reaction or the inordinately slow pace of geological change both require an intellectual stretch.

Experience is what transforms the unfamiliar into the familiar, and the conceptual framework in which early humans organized their experiences was not abstract. Survival demanded that they be keen observers of their immediate environment. For them it was literally a matter of life and death to stalk prey for food, and to avoid the reciprocal fate. A high awareness of the immediate environment was a life imperative. When early agriculture was becoming established, those who perceived local patterns of precipitation, the wet and dry seasons, and the seasonal oscillation of temperature through the year were better able to succeed in the production of food. But did early humans concern themselves with century-long trends in global temperature? Could they note, or even care, that this year's mean annual temperature was a tenth of a degree warmer than last year, particularly when the temperature changes by 20 or $30\,°F$ every twenty-four hours and even more seasonally? Or could they imagine that their cultivation of land would have an effect on the other side of the Earth or change the chemistry of the global atmosphere?

It is difficult for humans to focus on small incremental changes worldwide when big things are happening at home. The strategy of dealing with the immediate has served humans well when the greatest threats were local and looming large. For example, in a modern context, when humans are asked to consider the concept of global climate change, a phenomenon that is planetary in scope and which operates on a time scale that exceeds political term limits, generations and life spans, there is a hesitation, even skepticism, that arises because it is outside of the realm of ordinary experiences derived from day-to-day living. A caution emerges, a natural tendency to move into unfamiliar territory carefully. Uncertainty accompanies unfamiliarity.

GETTING OUR ATTENTION

Our senses are tuned to detect rapid change. When we are driving a car, we are alerted by the honking of a horn, the wailing of a siren, the sudden appearance of a brake light on the car ahead, a dog dashing into the street. All of these changes are registered against a backdrop that we term 'normal'. The horn and siren sound against a backdrop of ordinary continuous traffic noise; the brake light flashes where there was no illumination previously; the dog suddenly appears as an object of different size moving rapidly in a direction that crosses the flow of traffic. Each of these intrusions into the normal background grabs our attention, alerts us to possible hazards. (Conversely, when we are trying to fall asleep, and want to suppress alertness, we may turn to the background sounds of music or mindless talk from the radio or TV, or to specially designed sound tracks of waves breaking on the seashore or wind rustling the leaves.)

When an automobile accident occurs on the highway, the traffic following immediately slows and begins to back up. Similarly, when we approach a lane closure, and two lanes of traffic must narrow to funnel through only one lane, we notice the slow down. We notice that the commute to work takes a half-hour longer on a day when road repair crews are at work and lane closures occur. But do we notice the very slow extension of our daily commute caused by urban sprawl? Do we notice that the commute takes thirty minutes more today than it did a decade ago, when there were many fewer cars on the highway? The answer is probably no, because the change has come so slowly. The slowing effect of the gradually increasing traffic comes as an accumulation of very small incremental delays, not generally noticeable on a day-to-day basis. Our senses have not evolved to alert us to small incremental changes taking place over long time intervals. We develop an awareness of slow changes in the backdrop of our daily lives only through our personal memory, or through the collective memory that we call historical records. We are reminded of these changes when highway engineers report that a segment of the interstate highway system, envisioned and built fifteen years ago to accommodate 100,000 cars per day, is already carrying more than *twice* the number of cars it was designed for. Aha! So that's why it's taking me an extra half hour to get to work each day than when that highway opened.[1] The average American spent thirty-six hours sitting in traffic tie-ups in 1999, compared with only eleven hours in 1982.[2] People accept and adjust to the extension of their commute in part because it has developed incrementally over many years.

---

[1] How long-term trends creep unobtrusively into our lives is the subject of the book by Hal Kane, *Triumph of the Mundane: The Unseen Trends that Shape our Lives and Environment*, Island Press, 2001, 200 pp.

[2] Texas Transportation Institute, Texas A&M University, 2001. See also the *New York Times*, 8 and 9 May 2001.

CLIMATE IS WHAT YOU EXPECT, WEATHER IS WHAT YOU GET

Sensing climate change is equally difficult for the same reason. Climate is different than the weather. Weather is notable for its changes – yesterday high winds, today a thunderstorm, tomorrow sunshine. Because of the changeable nature of the weather on a day-to-day basis, we are alert to its variability and welcome the summaries of its likely course over the next few days presented to us in the daily weather report.

Climate, by comparison, is the long-term characterization of the 'average' weather. We describe the climate of a region with terms such as continental, mediterranean, or coastal maritime. Each of these terms implicitly carries a description of the annual average temperature, the seasonal temperature range, the annual amount of precipitation, the average number of days when the ground is covered with snow. When we are planning for a summer bicycling trip around Iceland, we consult a climate atlas to know what to pack. Such an atlas will tell us that we can expect the daily temperatures to be between 5 and 25 °C (41–77 °F), and there is a 30% probability of rain on any given day.

This type of information is climatological in that it portrays the long-term average conditions. Slow changes in climate are difficult to discern because they do not manifest themselves on a noticeable daily basis. Only when the average daily temperature is observed to trend upward or the annual precipitation downward over several decades do the climatologists, those keepers of the historical archives of the daily meteorological measurements, call to our attention that we are experiencing a change in climate.

There is also the temptation to interpret short-term departures from the norm as long-term trends. Many residents of the North American mid-continent will recall the summer of 1988 as a preview of hell. Record temperature levels were reached and sustained over long periods during that summer. Where I live in Michigan, set amidst the Great Lakes, residents endured more than forty consecutive days of temperatures exceeding 32 °C (90 °F) and ten days when the temperature exceeded 38 °C (100 °F). The grass was brown, the trees thirsty, the crops meager. The newspapers trumpeted that global warming had arrived with a vengeance. By contrast, in the summer of 2000, we had not a single summer day with the temperature as high as 32 °C, and rain came with great abundance. Obviously, one summer, whether it is in 1988 or 2000, does not determine a century-long trend, any more than the mean annual temperature at a single location determines the average temperature of the entire globe. We must be careful not to pay undue attention to short-term phenomena, at least in the context of identifying longer-term trends.

Long-term slow changes in the average annual temperature are particularly difficult for an individual to perceive, in part because these changes are typically small when compared with the temperature change that occurs between day and night, or between summer and winter. We are not equipped physiologically or psychologically to notice small slow changes superimposed on big rapid changes. The familiar phrase "Don't bother me with the small stuff" expresses our natural tendency to focus on the big things that are happening now, rather than the smaller changes taking place over long periods of time, even though the small incremental changes may ultimately have a big impact. Even a 'drop in the ocean', if repeated often enough, will raise sea level perceptibly.

In addition, living in the modern technological world generally means that, as far as temperature is concerned, we isolate ourselves from the changes taking place in the natural world. This isolation stems from the fact that many of us live and work in a 'climate-controlled' environment, inside buildings. In our thermal isolation, we set and adjust our machines to maintain a narrow temperature range indoors, and consequently we are out of touch with changes in the natural world. It is conceivable that the only way one might become aware of long-term trends in the average annual temperature would be through changes in the amount of energy consumed by the machines that control our indoor climate!

BEYOND THE HORIZON

If it is difficult for us to be aware of long-term changes taking place around us, it is equally difficult to develop a sense of what is happening elsewhere on Earth. In fact it is wholly human to imagine that what we experience locally is happening everywhere. If we have had a warm winter with little snowfall in Michigan, then it is easy to think that a similarly mild winter occurred everywhere. However, try telling that to the folks who live in Siberia, who during the same winter experienced bitter cold and record snowfall.

What we experience as individuals is the local expression of a global-scale process. The first-order pattern of regional climate variability on Earth is that it is warmer in the equatorial regions and colder in the polar regions, generally reflecting the variable amount of solar energy that each square meter of surface area receives. However, this fundamental pattern imposed by the Sun is altered by the hemispheric seasonal fluctuations that arise principally from the tilt of the Earth's rotation axis, and by the redistribution of Earth's heat by atmospheric circulation and ocean currents. The last are particularly influenced by the geography of the continents, which of course act to constrain and guide where the ocean water can go.

With such regional variability, some places being colder than average and others warmer, the concept of an average temperature for the globe becomes somewhat abstract (in the next chapter I describe how the global average temperature is obtained through aggregation of many local observations). And it is possible, even likely, that there is no place on Earth that actually experiences the global average temperature day by day. But this is no more abstract than the assertion that in a classroom of thirty students, one can calculate the average height and weight of the class, even though there may be no one in the class who has the average measurements.

Although the weather we experience is the local manifestation of the global meteorological system, we are not in the habit of thinking globally, despite the exhortations of environmentalists. However, the scope of our vision has also been significantly enhanced by means of Earth-orbiting satellites, enabling us to 'see' globally. A synoptic view of the weather patterns over an entire continent as seen from the eye of a satellite 22,000 miles above us can now be viewed daily on television, or on the World Wide Web. We can watch a hurricane develop in the Atlantic Ocean and make its way westward to a landfall somewhere on the eastern seaboard of America. The same satellite technology enables us to reach around the world with ease, speed, and clarity. Gone are the days when a call from Africa or South America to Europe or North America had to be booked a day or two in advance, and which, when finally put through, sounded little better than the tin-can-on-a-string technology of my youth. Cell phones, email, and TV signals beamed via satellite have brought instant communication to all parts of the globe. A submarine disaster in the Barents Sea, an illness at the South Pole, guerrilla warfare in the jungles of Colombia, or a newly erupting volcano pushing its cone above sea level in the South Pacific – all can now be viewed instantly in any corner of the planet.

Our geographic parochialism is also being overcome by world-wide economic activity engendered by the telecommunication revolution. Globalization is now a commonplace word used to describe international business and trading relationships. International corporations, governments, and workers recognize that what goes on in one part of the world affects all parts of the world. Currency instabilities in southeast Asia generate financial tsunamis on the bourses of the world; small differences in the pricing of a bond in London and in Hong Kong lead to large electronic transfers of capital in a split second. Reports on the meetings of international trade groups, and of the protests they engender, are front-page material, no longer buried deep within the business sections of the newspapers.

However, there are also subtle aspects of globalization that ironically make us *less* aware of the regional variability of economics and climate. In the early pre-history of humans on Earth, as agriculture became an important component of the human food supply, a prolonged drought would force human migration to new areas with sufficient precipitation to sustain agriculture. Members of the migrating community certainly were aware of the regional variability of the climate and of the stress that climate changes imposed on their lives. Today, in the globally integrated economy, diminished production in one region can be compensated for by increased yield arising somewhere else. To the consumer, a long way away from either the diminished or the enhanced production, the supply seems steady, without any concern for patterns of changing climate. To the typical consumer, it matters not whether the oranges come from Florida or Israel, as long as they are regularly on the shelf at the supermarket.

Globalization means more than tying the world together via telecommunications. It has sometimes been referred to as 'the end of borders'. Throughout much of Europe, passports are rarely needed, and many national currencies have been displaced by a single European currency, the euro. It also means an incessant flux of people and goods within and between the continents, leading to the globalization of unwanted hitch-hikers: the transport of viruses that bring diseases to new areas where they have never been encountered before. Exotic species such as the zebra mussel, which has found a new home in the Great Lakes of North America, arrived in the bilge waters of cargo ships from Europe. The outbreak of foot and mouth disease that ravished the farm animals of the British Isles in 2001 was imported via a circuitous route from Asia.

These environmental aspects of globalization, however, draw much less attention than do the economic aspects. The public is much less well informed about the global scale of environmental issues. While it seems straightforward for the public to grasp the idea of a global economy, it appears much more difficult for many people to think about a global environment. Why has there been so much uncertainty in the mind of the public about global climate change, or the global depletion of stratospheric ozone? Part of this disconnect, of course, arises from the deliberate efforts of vested interests to spread doubt and confusion about these issues. There is, however, a deeper reason for the skepticism. It is the notion that humans are impotent in the face of the vast forces of nature. Humans see their houses disappear in a flood or tornado, their communication and power distribution lines destroyed by an ice storm, their highways closed by a blizzard, their beaches disappear in a hurricane, their villages buried by landslides, their cities destroyed by an earthquake. It is difficult for a person to envision him- or herself as a powerful player in the global scheme of nature.

To an extent the individual is right, of course. He or she alone is not a big force of nature. Collectively, however, it is a far different story. Humans, now numbering over six billion on the planet, have a huge influence on the environment. The alteration of the chemistry of the atmosphere has been profound, through the introduction of ozone-depleting chemicals and climate-altering greenhouse gases. The pollution of the ocean with *sound*, from ships going somewhere around the clock, from off-shore oil rigs drilling and pumping, and with underwater transmitters generating strong signals designed to be 'heard round the world', has resulted in an astounding situation: no place in the ocean, no matter how remote, is free from the sounds created by human activity.

Humans as a team are leaving their mark on the global environment, with an efficiency at least as great as they have shown in integrating the globe through telecommunication and economics. However, we are lagging in recognizing our collective natural power to alter the environment on a global scale.

AT A SNAIL'S PACE

If century-long trends seem difficult for an individual to identify because an individual's life span ordinarily falls well short of a century, much longer-term trends might seem well nigh impossible to identify. Yet there are longer-term forces at work that play a role in Earth's climate and provide a backdrop of very slow change against which a significant trend over merely a century would seem like a piercing siren amidst the steady hum of traffic. The century-long trend would then become the event that catches our attention.

Longer-term factors that affect Earth's climate derive from variations in the shape of Earth's orbit about the Sun, and in the orientation of the Earth to the Sun. Earth's orbit around the Sun is almost but not quite circular, and the modest departure from a perfect circle renders the orbit into a slight ellipse, with the effect that Earth's distance to the Sun is not quite uniform over the year. When Earth is a little closer to the Sun, it receives more solar heating, and when a little farther, less. This variation in the amount of solar heating over the course of a year contributes a little to what we call seasonality on Earth (the more important factor in seasonality is the tilt of Earth's rotation axis). The significant aspect of the elliptical shape of Earth's orbit, at least in terms of long-term trends in the climatic regime, is that the shape of the ellipse is changing *very slowly* over time. The orbit about the Sun oscillates in shape, stretching, relaxing, stretching, relaxing, with a complete cycle taking a mere 100,000 years. And during this long period of orbital exercise, when the orbital stretch is at a maximum, the annual variation of the radiant heating is greatest; as the stretching relaxes, the annual oscillation of sunshine intensity diminishes, yielding a more uniform heating of Earth throughout the year.

While this 100,000 year dance of Earth and Sun is taking place, the Earth is also doing some rhythmics on its own. The axis about which Earth spins each day, the rotation axis, is today tilted away from a line perpendicular to the plane of Earth's orbit around the Sun. This tilt leads to the hemispheric oscillation of the seasons; if Earth stood upright and the tilt were absent, then the 'winter in Toronto when it's summer in Buenos Aires' phenomenon would not exist. But the tilt angle, today at about $23.5°$, is also slowly changing between $22°$ and $24°$, with a complete cycle occurring every 40,000 years or so, a slow-motion version of the back and forth bowing characteristic of practitioners of some of the world's orthodox religions. A bow to $24°$ accentuates the contrasts associated with the seasons, whereas a return to a $22°$ tilt diminishes the seasonality.

There is yet a third movement in this planetary exercise regime. The spin axis, bowing and rising every 40,000 years, is also doing a 25,000 year pirouette similar to the slow motion wobble of a spinning top. The effect of this motion is to slowly alter the timing of the seasons within the year. Whereas today the northern hemisphere is tilted toward the Sun in June, July and August and North Americans, Europeans, and Asians enjoy summer, in only 12,500 years those will be the winter months in the northern hemisphere.

All three of these long-term factors – the periodic stretching of Earth's orbit, the change in the angle and the change in the orientation of the spin axis – together make up a kind of climatological Tai-Chi that impacts the amount of heat that Earth receives from the Sun and how that heat is distributed geographically and seasonally over the face of the planet. These periodic changes are slow, taking tens of thousands of years to complete a cycle. In such a context, a significant climatic change over a single century, such as that caused by the human enhancement of the greenhouse effect in the twentieth century, can be seen as very rapid indeed.

Even slower changes shape Earth and its climate, changes that occur not over tens of thousands of years, but over tens of *millions* of years. In the context of such very slow changes, things that happen in just tens of thousands of years seem rapid! The difference in pace is exactly the same as between a second and a thousandth of a second. Photographers will appreciate that difference when applied to exposure times. An exposure at 1/1000 second captures a sharp image of an athlete in motion, whereas a one second exposure would yield only a blur. Processes we described as Tai-Chi only a moment ago would against a background of even slower changes appear as lightning fast martial arts.

REARRANGING THE FURNITURE

What are the real snail's pace processes that affect the climate on Earth? Plate tectonics and continental drift. The geography of Earth's surface is constantly being rearranged as the fragmented outer shell of rigid rock that forms the solid surface moves slowly about, in response to very large-scale movements in the deeper interior. The speed at which these giant mosaic tiles passively ride atop the interior currents is just an inch or two (a few centimeters) per year, approximately the speed at which fingernails grow. Today the Americas are distancing themselves from Europe and Africa at that slow pace, and the Atlantic Ocean is growing wider, year-by-year, century-by-century, millennium-by-millennium. If we looked at the map ten million years ago, at an earlier stage of this slow separation, the Atlantic would be some 500 kilometers narrower than today. And 200 million years ago, there was no Atlantic Ocean at all. One could walk from South America to Africa without touching salt water.

The patterns of ocean currents depend on the topographic configuration of the Earth's surface. The oceans occupy the low places; the continents stand above it all. The location of the continents determines where ocean currents can and cannot go. The last major change to the geography of the continents, at least with respect to ocean circulation, came about three million years ago, when the Isthmus of Panama was uplifted, cutting off the connection between the Atlantic and Pacific Oceans. No longer could there be an east–west circulation in the equatorial region, and the flow in the Atlantic adjusted to become principally a north–south pattern, with a northward flow of warm water, the Gulf Stream, occurring at the surface, and a southward flow of colder water at the bottom of the ocean; together they form a gigantic fluid conveyor belt whose principal cargo at the surface is heat.

Because of the Gulf Stream in the Atlantic, which flows northward across the equator to high northern latitudes, the climate of maritime Canada, southern Greenland, Iceland, Britain, and Norway is much milder than other places in the northern hemisphere such as central Canada or Siberia, which sit equally far north but far from the sea. Even Murmansk, the Russian port situated well north of the Arctic Circle, remains ice-free year round because the warm waters of the Gulf Stream curl around the northernmost point of Norway into the Barents Sea. Were the patterns of ocean circulation to change, climate change would inexorably follow.

There have been other slowly evolving reconfigurations of the geography that were accompanied by large climatic adjustments that proceeded apace. In the southern hemisphere, the opening of a continuous oceanic path around Antarctica some thirty million years ago allowed a west to east circulation to develop. That oceanic circulation provided a barrier to warmer waters penetrating to the southern high latitudes and effectively isolated Antarctica climatologically. The accumulation of the Antarctic ice cap followed, and it has persisted to the present day. While such climate changes are fascinating to geologists, they have only marginal relevance to assessing the significance of the climate changes of the twentieth century. These recent changes must be assessed against the background characteristics of the global climate system in the geologically recent past, a time interval in which the general patterns of oceanic circulation took on modern form. That time interval is essentially the past three million years, since the closure of the Isthmus of Panama.

The past three million years can, in some ways, be seen as a 'broken record'. Layer upon layer of sediment accumulated on the ocean floor in that time interval, containing shells of microscopic marine organisms. These shells tell a repetitive story of some thirty cycles of large-scale continental glaciation. In ways not fully understood, the oceanic circulation pattern imposed by the closure of Panama set the stage for cyclical glaciation. But the record of repeated glaciations found in the sediment layers at the bottom of the ocean is unambiguous.

How do sediments in the ocean tell a story about glaciations? The concentrations of various chemicals in the ocean depends on how much water is in the ocean – more water leads to chemical dilution, less water to greater concentration. The chemistry of an organism's shell reflects the chemistry of the seawater at the time the organism was living. The water that gets frozen into continental ice sheets comes from the oceans, evaporated, transported, and dropped as snow on the continents. When there is much ice on the continents, there is less water in the oceans, and vice versa. The most recent glaciation, the ice advance that reached a maximum extent some 20,000 years ago, is simply the last stanza of this repetitive chorus. Since then, the ice has been melting back, albeit not always in a smooth and steady retreat.

The geological record surely has some lessons for us in understanding the natural variability of Earth's climate, but the pathways through the past provide ample opportunities for misinterpretation as well. The further back in time we try to see, the murkier the picture becomes. Observations become fewer and more irregularly distributed around the globe, the ability to resolve short intervals of time becomes more difficult, and the determination of how long ago something happened becomes less precise. In discussing climate changes over just the past 2000 years, there are some well-entrenched climatological concepts in the scientific literature that deserve re-evaluation. Introductory textbooks on climatology will mention the Medieval Warm Period (MWP; ca. 800–1200 AD) and the Little Ice Age (LIA; ca. 1300–1850 AD) as global climatic excursions. The MWP is the time interval when the Vikings established settlements in Iceland and Greenland, and culture and trade expanded in Europe. Conversely, the LIA was a time of climatic deterioration, a time when the Vikings abandoned their settlements in Greenland, the area of winter sea-ice around Iceland expanded substantially, mountain glaciers in Europe advanced, and agricultural production declined.[3]

Just as we may be tempted to think that what we experience locally is characteristic of the entire globe, it is easy to slip into that same illogic when considering evidence from the past. Some scientists now question whether the MWP and the LIA were truly global in extent. The questioning arises from the Eurocentric nature of the above-mentioned evidence. Because so many of the historic records are European, there is the possibility that the MWP and LIA were only regional climatic phenomena affecting principally the North Atlantic region, with Europeans noting and recording the effects. Today, we know much more about the sensitivity of climate in the North Atlantic and its close coupling to what goes on in the Arctic Ocean. The Arctic Ocean has no effective link to the other oceans of

---

[3] The book by Brian Fagan, *The Little Ice Age: How Climate Made History 1300–1850* Basic Books, New York, 2000 (246 pp.) provides fascinating detail about day-to-day life during the Little Ice Age.

the world except through the far North Atlantic, where changes in the Arctic can have dramatic effects in the lands bordering the North Atlantic. But do those effects, so well documented in the North Atlantic region, constitute a global climate shift? Or are they good examples of significant regional climatic events that have been misinterpreted as global?

These are not easy questions to answer. Even if the MWP and the LIA have their genesis and principal effects in the North Atlantic region, the changes there can be exported to other parts of the globe, probably in muted form, through atmospheric and oceanic circulation patterns. Considerable uncertainty still clouds the question of how climate change in one part of the globe is linked to changes elsewhere. However, we must be cautious in making global interpretations from geographically limited observations. A global story generally requires global observations.

CAN SMALL CHANGES HAVE BIG IMPACTS?
In addition to the difficulty humans have in detecting changes taking place slowly over large regions of the globe, there is yet another source of skepticism about the importance of global climate change: the size of the change experienced or anticipated. When climatologists point out that Earth's annual average temperature increased by about $1\,^{\circ}C$ (almost $2\,^{\circ}F$) over the twentieth century, or that it will increase by another few degrees over the twenty-first century, it is fair to ask whether such a change is important. After all, the temperature fluctuates much more than that from day to night and from season to season.

Part of the answer lies in the fact that rather large short-term fluctuations usually can be accommodated without serious damage to living things. Physiological mechanisms have evolved that enable plants and animals, large and small, to adjust to, or at least tolerate, short-term extremes as long as tomorrow or next month or next year brings some relief. Thus the various components of Earth's biota can sustain a day without food, or a week of extreme cold, or a year of low precipitation. But we can all envision the consequences of a year without food, or a decade of drought. The protective mechanisms that biological systems have developed do not have infinite elasticity. They can be stretched to the breaking point even if the disturbance is small, as the small effects accumulate over time. We have all heard the adage of 'the straw that breaks the camel's back', where the final increment of load on the camel is very small, but the cumulative effect is catastrophic. There is a threshold, a load limit, which when exceeded leads to a system breakdown, no matter how slowly or incrementally the limit has been approached. I will return to this topic in a later chapter when we take up the question of whether the global climate system has the potential of being loaded to some breaking point.

Many complex systems are very sensitive to small changes. Most readers will be familiar with the AM and FM radio bands, where while driving to and from work they tune in to their favorite news report, talk show or sporting event. I dial regularly into the 89.1 megahertz frequency, which is one of several public radio stations serving my community. The frequency designation means that the station signal is made up of electromagnetic waves that arrive with a frequency of 89.1 million every second. However, with just a slight turn of the tuning dial to 89.2 megahertz, National Public Radio is replaced by static. And at 89.7 classical music appears. A small change in the millions of waves the radio is receiving means a great deal, because the radio is built to be sensitive to small changes.

The human body provides an example of a complex system built to run at a certain temperature, namely 37 °C (98.6 °F). If our internal temperature rises by only a degree or so, it is an indication that we are ill, that the system is encountering some problems. One degree may not seem like much in the abstract, but for a finely tuned machine like the human body, with temperature-sensitive organs and a delicately crafted thermostat, a small deviation is not a trivial matter.

Nor is a small change a trivial matter to the global climate system, another complex system of coupled and intertwined processes that absorb or reflect sunshine, that transport heat around the globe through the atmospheric and oceanic circulatory systems, and that take and return chemicals to and from different parts of the system. Tamper with one part of the system and the effects spread widely throughout the rest of the system. In the delicately balanced global climate system, one degree means a lot if it is both global and of long duration.

AN EVOLUTIONARY SPECULATION

Many of the themes of this chapter have also been developed in a recent book by the noted biologist Edward O. Wilson. In *The Future of Life*[4] he writes:

> The human brain evidently evolved to commit itself emotionally only to a small piece of geography, a limited band of kinsmen, and two or three generations into the future. To look neither far ahead nor far afield is elemental in a Darwinian sense. . . . Why do [we] think in this shortsighted way? The reason is simple: it is a hardwired part of our Paleolithic heritage.

If indeed Wilson is correct that evolution has hardwired spatial and temporal parochialism into our brains, we might wonder, in a larger context, why nature has chosen not to endow every new generation with more of the accumulated knowledge of previous generations. The philosopher George Santayana advised, "Those who cannot remember the past are condemned to repeat it."[5] Why must children of every generation go to school to learn elementary arithmetic? Why can't we be born with the multiplication tables hard-wired into our brains? Why have centuries of science and engineering students relearned the differential and integral calculus that Isaac Newton and Gottfried Leibniz invented in the seventeenth century? Would it not be more efficient, and have little if any downside, to have children born with these powerful and unchanging tools, rather than struggling to acquire

---

[4]Published in 2002 by Alfred A. Knopf, a division of Random House, Inc.
[5]George Santayana, *The Life of Reason* (vol. 1), 1905.

them time and time again? What are the advantages of an evolutionary strategy that discards so much with the passing of mature individuals, only to re-instill much of that acquired knowledge in the generations that will follow?

Nature of course does hard-wire us with the true fundamentals. Our genes endow us with a body that provides mobility, cognitive tools with which we acquire information, and a brain that acts as a control center and database. Almost everyone rolls off the assembly line with these essentials. And then the learning begins. As noted earlier, children come into the world naturally curious. With their cognitive tools, they look, they touch, they listen, they taste. Within their brains, they sort, compare, and evaluate these experiences. Eventually they go to school to learn their multiplication tables, and a few even to learn calculus and differential equations.

The price we pay by not hard-wiring arithmetic into our brains is that we, each individual and each generation, must learn to add, subtract, multiply, and divide anew. But could it be that forcing each new human being to learn about the world is nature's way of generating creativity in each new generation? Perhaps nature has chosen not to bias each new generation with too much of the conventional wisdom of the past. The world is always changing, and humans, like it or not, must also change. The ways in which they confront and accommodate future change often will not be found in the conventional wisdom of the past. This is the dilemma that Wilson defines; the lessons of the past have become impediments that make our adaptation to the future difficult.

Has nature done us a favor in not giving us much baggage to carry that may stifle creativity and be ill-suited to guiding us smoothly into the future? Might it be possible that having our familiar 'base 10' arithmetic hard-wired into our brains would have impeded the conceptualization of 'base 2' (binary) arithmetic, which is the underpinning of modern digital computers? Try as we may, we cannot make the world static. Change is the norm, and perhaps the ability to recognize and adapt to change is a more important skill in human evolution than other skills that focus on conventional wisdom and preserve the *status quo.*

In this chapter, we have seen that phenomena such as global warming are very difficult concepts to evaluate on the basis of personal experience. We have a good understanding of what happens around us in the short run, but have a harder time with long-term trends in the average temperature over the entire globe. The temporal and spatial scales of long-term global phenomena are simply beyond our individual sensory capabilities. We can deduce such phenomena only through collective observations and shared information, from around the world and over centuries. To accept and have confidence in these distillations of collective experience requires that we go beyond the familiar base of our own personal experience, that we step out into unfamiliar and uncertain terrain. It is only natural that such steps are taken cautiously. And it is only natural that broad generalizations about an unfamiliar world meet with some skepticism.

In the next chapter, we take up the uncertainties of measurement and data aggregation, topics that science students face on their very first foray into a laboratory. Someone once said, "If you can't measure it, it isn't science". But what are the tools for measurement, and what uncertainties are associated with them? And after we have some measurements, what do we do with them? This part of the garden of uncertainty is an older, well-established plot that accordingly offers some interesting historical perspectives with contemporary relevance.

5 Fever or chill?

We're trying to measure bacteria with a yardstick.

Professor John A. Paulos,[1] Temple University

At a fundamental level, scientific uncertainty begins when we make measurements. What do we use to make a measurement? How well can that tool accomplish a measurement? To what precision can we determine the size or mass or temperature of an object? If we repeat a measurement many times, how closely will the individual measurements agree with each other?

Professor Paulos' comment about measuring bacteria was made in an unusual context that I will tell you more about later. The remark underscores, however, the importance of selecting a measuring device appropriate to the task at hand. One does not need the experience of a laboratory scientist to recognize that the likelihood of obtaining an accurate measurement of the length of a bacterium using a yardstick is intrinsically low. The smallest subdivision of the yardstick, usually 1/16 of an inch, a little less than two millimeters, is so much greater than the dimension of a bacterium (actually about 10,000 times greater) that, on the one hand, one cannot say much more about the length of a bacterium other than it is very much smaller than 1/16 of an inch. On the other hand, a yardstick could, in principle, estimate the length of five million bacteria lined up end to end.

Most textbooks of physics or chemistry introduce uncertainty in the context of measurement. These discussions typically center on the sensitivity (or precision) and accuracy of the measuring device, and the repeatability of a measurement of the same object or condition. Sensitivity relates to an instrument's response to change, to how small a difference in a quantity the measuring instrument can detect or sense. Accuracy describes how closely the measured value

---

[1] *New York Times*, 22 November 2000, p. A31.

is to the true value. A thermometer that produces a small but noticeable change in the height of the liquid column for a tenth of a degree change of temperature, but not for a hundredth, is said to have a sensitivity of one tenth of a degree. Consider, as an example, the temperature of boiling water, which in the Celsius scale is defined to be exactly 100 degrees. A thermometer that registers 99.7 or 100.3 °C in boiling water is said to have an accuracy, or margin of error, of 0.3 degrees. If in a particular experiment, a chemist did not need to know the actual temperature any closer than within a half degree, a less-accurate (and much cheaper) thermometer would suffice.

If you measure the same thing twice, or ten times, intuitively you believe that the measurement should be the same each time. Yet, whether it is a measurement of the length of a piece of carpet or the temperature as displayed by a thermometer, every time one reads the tape-measure or the thermometer there is the task of estimating the reading when it falls between the smallest subdivisions of the measuring device. Depending on the angle of viewing and the experience of the viewer, the estimate will likely differ from measurement to measurement. Similarly, if one counts the number of white blood cells in a blood sample, in principle one might think the count should be the same each time no matter which laboratory technician performs the count. However, in practice, the count may differ from one technician to another, and even for the same technician from one time to another. In any collection of repeated measurements, there will be a range of values. Which one, if any, is correct? Here we often resort to the power of statistics to determine an average, and to estimate the probability that the average we calculate from the collection of measurements is the true value. That probability is usually higher if we make many measurements, and lower with only a few measurements. But the probability will never be so high that we can be *absolutely* certain of the result. Some uncertainty remains.

These same issues of repeatability of course arise when different people make measurements of the same object, but use different measuring instruments or different techniques. Thus the speed of light has some uncertainty associated with it because of the different experimental approaches used to determine this important quantity. The uncertainty may diminish over time as more sensitive measuring instruments are conceived and utilized.

Repeatability is a fundamental tenet of science. If one investigator announces an astounding scientific discovery, many others will try to confirm the phenomenon by repeating the experiment or measurement. In 1989, when a procedure to fuse atoms at low temperature[2] (popularly referred to as 'cold fusion') was presented to the world, an extraordinary excitement ensued. If cold fusion actually proved possible, it held the promise of virtually limitless inexpensive energy. Researchers everywhere rushed to their laboratories to try the experiment for themselves. In this case, repeatability proved elusive, and the prospect of cold fusion fell back to Earth like a spent roman candle. The more recent claim of fusion accompanying the implosion of bubbles, mentioned in Chapter 2, will surely be tested quickly in laboratories around the world.

Not all measurements can be repeated, because the quantity being measured is a moving target. A thermometer outside your window will indicate the temperature is changing all the time. People who purchase older houses discover that the angles between walls, floors and ceilings depart from their original ninety degrees because of settlement and sagging over time. From day to day the geometry may appear static, but over the years changes accumulate. The height of Mt. Everest is changing as the crust of the Earth is pushed upward by tectonic forces and reduced by erosion. We can obtain a sequence of measurements over time, but because time marches on, we never have the opportunity to repeat many earlier measurements.

---

[2]Fleischmann, M., Pons, S., and Hawkins, M., *Journal of Electroanalytical Chemistry*, vol. 261, p. 301, 1989.

Another aspect of inaccuracy, and therefore uncertainty, in any measurement relates to calibration of the measuring device. How do we know that the markings on a tape measure or yardstick or thermometer are properly spaced? Can we be certain that a distance of one foot on these measuring devices is actually one foot? Can we be certain that when a fever thermometer reads $98.6\,°F$, it is actually that temperature, or might it be $98.5\,°F$, or even worse, might it actually be 102? How does the manufacturer of the thermometer determine what height of the liquid column actually corresponds to 98, 99, and $100\,°F$? The answer is that they use another 'better' thermometer to tell when the temperature of an object or a fluid is actually at 98 and then put an appropriate mark on the glass at the level of the liquid in the thermometer being calibrated.

But how did the 'better' thermometer get its markings? Here one quickly recognizes the need for a standard: something we *assume* to be exactly one meter long or exactly at some specified temperature. In the USA, the final say on such matters resides with the National Institute of Standards and Technology. The performance of any measuring instrument can be compared with the accepted standard value; those instruments that yield measurements that are very close to the standard are 'better' instruments (and usually cost more) than those that yield less-consistent results. With any measurement, however, aside from the uncertainty of reading the instrument, there is always an uncertainty of some magnitude about whether the instrument is measuring the true value.

STANDARD MEASURES

John Quincy Adams, Sixth President of the United States (1825–1829), said:

> Weights and measures may be ranked among the necessities of life to every individual of human society. They enter into the economical arrangements and daily concerns of every family.

Standards have been established for measurements of distance, time, temperature, mass, and many other physical quantities. And these standards continually evolve over time, as new technical methods are developed. The history of the standard meter, the fundamental unit of length in the metric system of measurements, provides a fascinating insight into how quantities are defined and standards established.

The story of the standard meter begins in the early eighteenth century. At that time, the scientific world was still marveling at the insights developed by Isaac Newton half a century earlier in the science of mechanics. In particular, his success in describing the motions of the planets about the Sun, in terms of his simple laws of motion under the influence of gravitational forces, brought great acclaim. But his laws of mechanics were not confined to explaining planetary orbits. They were relevant to many other natural phenomena as well, among them the shape of the Earth. The fundamental role that gravity plays in this topic is that it will try to tuck all the mass of the Earth into as small a volume as possible, with every piece of the Earth packed as close as possible to the center of gravity. The outcome is that Earth and the other planets would ideally be shaped by such a process into spheres.

However, if the sphere is also rotating about an axis, as all the planets do, another force, the centrifugal force of rotation, comes into play, and a planetary figure must adjust to this additional force. Newtonian mechanics predicted that the spinning of the planet would cause it to bulge outward a little at the equator and depress slightly at both poles. The shape at equilibrium with both gravitational and centrifugal forces would then be an ellipsoid, with its long axis in the equatorial plane and the short axis passing through the poles.

Scientific theories always undergo tests and probes to see if they are incomplete or have weaknesses, and Newtonian mechanics was no exception. A direct measurement of this predicted shape of the spinning Earth could then serve as another confirmation or verification of

Newtonian mechanics. Accordingly, in 1734, less than a decade after Newton's death, the French Academy of Science conceived an experiment that would provide a crucial test for the revolutionary physics proposed by Newton. The French sent geodetic scientists and surveyors to northern Scandinavia and to Peru to measure the curvature of the Earth in these areas that were, respectively, near the pole and near the equator. Three years of field surveying confirmed that indeed the Earth was bulging at low latitudes and flattened near the pole, and Newtonian physics had passed another important test.

So what is the connection of this vignette to establishing standards of measurement and, in particular, to defining the standard of length we call the meter? As an outgrowth of this geodetic expedition, the French Academy of Science made a decision to use the Earth itself, rather than body parts of some distinguished monarch, as a measurement standard. They determined the distance from the equator to the pole along a meridian, divided it into ten million equal parts, and called the length of that unit one meter. Thus the meter as a unit of measurement of length was originally defined as 1/10,000,000 of the distance from the equator to the pole along a meridian. Of course they did not actually lay out a long tape measure from the equator to the pole. They knew that that distance corresponded to $90°$ of latitude, and if they could carefully survey the length of $1°$ of latitude at both high and low latitudes, then simple scaling could extend their result to the full distance. With the adoption of this new standard, no longer would science be dependent on the length of King Somebody's foot, or other similarly 'unscientific' dimensions. Scientists thereafter had a measuring standard tied directly to the enduring Earth, not one tied to a mortal human's dimension. Eventually this geodetic determination was translated into markings on a bar of platinum–iridium alloy, kept at a standard and invariant pressure and temperature to avoid slight changes of dimension owing to changing environmental conditions. This metallic bar was housed in Paris and by international agreement served as the ultimate physical reference as to exactly how long a meter was.

However, over the years, geodetic science improved, and better measurements of the Earth's dimensions led to departures from what the French had determined in 1734. Today, instead of exactly ten million meters between the equator and the pole, geodesists have determined that there are 10,002,286 meters along such a meridian. The error in the original determination of 1734 was only about two parts in ten thousand, or about 0.2 millimeters for every meter. That was quite an accurate measurement for its time, amounting to an error of only a hair's width in every meter. Wisely, we have not elected to redefine the meter in terms of this newer and more accurate measurement of Earth's geometry. Besides, as the Earth's rotation slows over geologic time, the bulge of the equator and the flattening of the poles will gradually and continually adjust, making the Earth's dimension an ever-changing quantity, not a good characteristic for an international measurement standard.

Recent definitions of the meter have in fact been totally divorced from Earth's geometry in favor of something that is even more stable and unvarying. In 1960, the meter was redefined in terms of a specific number of wavelengths (1,650,763.73 to be exact) of a particular spectral line in the electromagnetic radiation emanating from an excited atom of krypton-86. No longer did the meter have any direct tie to the geometry of Earth.[3] However, the 1960 redefinition was to be short-lived. In 1983, the meter was again redefined, this time as the distance traveled by light in a vacuum during the time interval of 1/299,792,458 of a second. This definition is effectively a statement that the speed of light in a vacuum is 299,792,458 meters per second, and an implicit recognition that the standard meter cannot have an accuracy greater than that which characterizes the speed of light.[4]

Similar stories can be told for each of the other fundamental units in our system of measurement. The standard for mass in the metric system, as originally defined by the French, was called the

---

[3] It is interesting to note that the word 'geometry' itself literally means 'Earth measurement'.

[4] See the discussion by Daniel Kleppner in *Physics Today*, March 2001, pp.11–12.

gram, a quantity equal to the mass of one cubic centimeter of water at a specified temperature. However this standard was physically a very small quantity, about a thimble-full of water, and difficult to measure precisely. Today the mass standard is the kilogram (1000 grams) and is represented by a cylinder of platinum–iridium alloy kept in a vault in the International Bureau of Weights and Measures in Paris (a duplicate cylinder is housed in the National Institute of Standards and Technology in the USA). This one kilogram cylinder, the mass standard, is the only standard remaining in which the quantity is represented by an actual physical artifact. The standard for time, the second, was originally defined as 1/86,400 of the time it took the Earth to turn once on its axis, one day comprising twenty-four sixty minute hours, with each minute comprising sixty seconds. However, detailed modern measurements of the Earth's rotation have shown it not to be uniform and, therefore, unreliable as a standard of time. Today, the second is officially defined as the duration of 9,192,631,770 cycles of a certain wavelength of radiation from the cesium-133 atom. As with the meter, the definition of the second has been divorced from a characteristic of the Earth and has been replaced with the much more stable behavior of a particular atom.

LOST AT SEA

Necessity may be the mother of invention, but uncertainty is surely her twin. Uncertainty has long been a stimulus of creativity, not in the least with regards to chronometry, the measurement of time. One of the commonest uncertainties expressed by someone situated in unfamiliar surroundings is "Where am I?" This age-old uncertainty is felt particularly by mariners at sea, out of sight of familiar landmarks to orientate them. The entire art and science of navigation arose from the desire to know one's location on the face of the Earth. Without landmarks, sailors used what was available, and those were 'skymarks', the stars and planets above. But because Earth rotates on its axis daily, the celestial guides do not stay put in the sky but appear to move as Earth rotates beneath them. There is one exception, however, the star called Polaris or the North Star. Polaris is situated directly above the North

Pole, along the extension of Earth's rotation axis. It appears stationary in the nighttime sky, and a simple measurement of the angle that Polaris makes with the horizon gives observers the latitude of their position. If you are at the North Pole, Polaris is straight above you, or $90°$ up from the horizon – and that is the latitude of the North Pole, $90°$ north. Seafarers in the northern hemisphere long ago recognized the simplicity of determining latitude from Polaris.

Longitude, however, was another story. Longitude is a measure of how far east or west one is from an arbitrary starting point. Today, by international agreement, we define zero longitude as the meridian that passes through Greenwich, England, the so-called prime meridian. Historically, however, there were other candidates, and for many years the French measured longitude from the meridian passing through Paris. The problem of determining the longitude of a position essentially amounts to a careful measurement of the time when a navigator measured the angle from the horizontal upward to a star that was moving slowly across the sky as the Earth rotated beneath it. The Sun is a good example – it rises, slowly arcs upward to reach a high point, and then begins its descent to sunset. The time of sunrise, zenith, or sunset (or for that matter any other intermediate position), and knowing what day of the year it is, uniquely determines how far east or west of the prime meridian the observer is. The trick is merely to have a good clock from which to read the time when the observation is made.

In the early days of exploration, most clocks were mechanical devices with gears and springs. They were imperfect machines that typically gained or lost time at rates that were highly variable. The drift of a mechanical clock may depend on the local temperature, the atmospheric pressure and relative humidity, the nature of the lubricating oil, how tightly its spring is wound, whether it is mounted horizontally or vertically, and more. For ships at sea, underway for months without a landfall, the ability to locate the ship accurately on the map, or to place newly discovered lands or dangerous reefs on charts, was linked to the ability to keep time accurately. And so the challenge was to design and build a stable and accurate marine chronometer in order to narrow the uncertainty about where you were on the open sea. The British Admiralty had a keen interest in the development of such an instrument and offered an extraordinarily rich prize to be awarded to the inventor of a chronometer that met Admiralty specifications. The story of the ensuing competition among clockmakers of the eighteenth century to construct this timepiece, and of the later foot-dragging by the Admiralty to pay the prize to the winner, is the subject of a fascinating book by Dava Sobel.[5]

"THE EAGLE HAS LANDED"

Probably no radio transmission has been received with such excitement as the clear and unmistakable words voiced by astronaut Neil Armstrong, announcing the arrival of the Apollo 11 spacecraft on the surface of the Moon: "The eagle has landed". In the early days of radio communication, however, the radio transmitters and receivers were crude devices by today's standards. A comment that anyone sending a message to a distant receiver liked to hear was that he was coming in "loud and clear", instead of "weak and garbled". For a signal to be received as loud and clear, it must not be masked by other sounds such as static, crackling, hissing, or humming, nor should the signal be so distorted in the transmission as to be undecipherable when received. When other sources of sound interfere with the reception of the transmission of interest, we say we have a 'noisy' situation in which it is difficult to hear the 'signal'. The same description could apply to a wedding reception where the dance music and the chatter of other conversations make your own voice difficult to hear above the din. The signal is lost in the noise. Conversely, the sound of the police siren or the bleating of an ambulance klaxon is purposeful; they are designed to be signals that easily stand out from ordinary traffic noise.

Laboratory scientific protocols and scientific measuring instruments are like a radio receiver in that the scientist wants to measure

---

[5]Dava Sobel, *Longitude: The True Story of a Lone Genius Who Solved the Greatest Scientific Problem of His Time*, Walker and Company, New York, 184 pp., 1995.

only the property of interest and to minimize irrelevant observations that arise from other sources. It is not possible to avoid all noise, and therefore at some sufficiently small level of signal, the whisper will be lost in the noise. Consequently, in designing a scientific instrument, it is important to be able to estimate the size of both the signal and the noise, so that the instrument will be up to the task in terms of sensitivity and accuracy.

Noise, now used in the broad context of anything that obscures a signal, arises in many ways. In the field of dendroclimatology (the study of tree-rings to reveal aspects of the local climate during the lifetime of the tree), the scientists analyze many trees in a region, rather than just a single tree. They do this because any single tree may display significant local noise that is specific to the location of the tree. A tree located in a well-watered valley may be less vulnerable to fluctuations in precipitation than another tree in a high remote nook of the drainage basin. A tree in a forest may suffer competition for sunshine whereas sparsely distributed trees may not. One tree's roots may have had its access to water impeded by an earthquake, whereas just a few miles away the quake had no effect. In short, no individual tree can be trusted to tell a reliable story. However, an ensemble of many trees from a region, each influenced by site-specific effects and by changes in the regional climate, enables the scientists to separate the regional signal from the site-specific noise.

COUNTING VOTES

Professor Paulos' comment about trying to measure bacteria with a yardstick was not made in the context of measuring bacteria in a laboratory, but rather in the context of determining the outcome of a close election. In November of the year 2000, Americans were drawn into a prolonged interval of uncertainty as to who had been elected President of the United States.

In most democratic elections, the task of determining who has won is conceptually simple: count the votes, and the candidate with the greater number of votes is the winner. In the state of Florida, which proved to be the pivotal state in determining who would become president, the initial count of nearly six million votes was almost evenly divided between the two principal candidates, George W. Bush and Al Gore. The initial difference in the vote totals between these two candidates was less than 1,000 votes, which means that out of every 6,000 votes cast, the vote totals of each candidate would differ by only a single vote. Naturally, with so much at stake, the apparent runner-up asked for a recount to ensure that no mistakes had been made in counting the ballots.

Anyone who has ever attempted to count a large number of items, say the number of coins in a large jar, knows that getting a fully accurate count is not as straightforward a task as it may seem. As an example, let us imagine a barrel filled with 10,000 green and blue marbles, and we want to determine how many of each color are in the barrel (a simulated tally of ballots in a two-candidate race). If we empty the marbles onto the floor and separate and then count them by color, there are many opportunities for an uncertain result to emerge. Through inattention, distraction, misidentification, or arithmetical errors in tallying, the final counts of blues and greens may be different each time a different person undertakes the count.

If the goal is merely to determine whether there are more greens than blues, usually the differences between the several counting results are of no consequence. If in the collection of 10,000 marbles there are actually 6,000 blues and 4,000 greens, it does not matter if one person tallies 5,984 blues and 4,016 greens, and another counter determines there are 6,007 blues and 3,993 greens. Nor does it make a difference if a third counter records 5,991 blues and 4,007 greens, which together add up only to 9,998 marbles (perhaps two rolled under a nearby cabinet and were overlooked). Nor is there alarm with a result that yielded *more* than 10,000 marbles (perhaps a few marbles broke into two pieces, and each piece was counted separately). There might even have been some difficulty in telling whether a marble is blue or green, owing to variability in the dye during manufacture, or perhaps someone counting them was marginally colorblind.

None of these uncertainties cast a doubt on the principal result that there are clearly more blues than greens; Candidate Blue has won the election, and Candidate Green has lost. That is because the range of uncertainty is much smaller than the difference between the candidates' vote totals. Let us assess the results of the count as accurate to plus or minus ten votes (this means that the final tallies could be in error by as much as ten votes up or ten down). With a difference of some 2,000 votes between candidates Green and Blue, the twenty-vote range of uncertainty is only 1% of the difference. The signal is clearly much greater than the obscuring noise, and the uncertainty in the result is obviously insignificant. The signal is loud and clear: Blue has won. However, if the marbles in the barrel were in fact evenly divided in color, 5,000 blue and 5,000 green, then a twenty marble range of uncertainty in the counting could easily yield a different winner in every recount. The measuring instrument, that is, the ensemble of procedures used in the marble count, simply did not have the necessary accuracy to determine that blue and green were equal.

In the 2000 United States presidential election, the candidates' respective totals in Florida were separated by only a few hundred votes out of almost six million cast. A principal source of uncertainty as to whether a ballot should be included in the final tally centered on a voting procedure in which a voter selected a candidate by punching out a small piece of stiff paper, called a chad, from a ballot card that listed all of the candidates. Some chads were tenacious, however, and did not fully detach. Should a chad left hanging from a single attachment point be counted? Or from two or three attachment points? Or in cases where no detachment occurred at all, but the chad was indented, perhaps indicating that the voter had intended to vote for that candidate but was simply unsuccessful in separating the chad from the ballot with the tool provided?

Other procedural aspects clouded the validity of some absentee ballots actually cast. Was the application for an absentee ballot properly filled out? Was the absentee ballot mailed and received by the calendar deadline? Some absentee ballots were misplaced but later found. Other ballots of different design were confusing to some voters and led to votes being cast inadvertently for a different candidate, not the person the voters thought they were selecting. And some few votes were probably cast by ineligible voters, some unregistered, some non-citizens, and perhaps some even voting twice in different precincts. There were also allegations that eligible voters were prevented from voting because their names had been erroneously purged from the list of registered voters.

Flaws in the execution and tallying of the vote in Florida are not unique to Florida. In some measure, and to varying degree, they occur in elections everywhere. They are intrinsic to the election process. The saving grace is that usually these flaws are of no consequence. The measuring instrument that we use to determine winners in an election is usually accurate enough for the task. But in a very close election, the inherent inaccuracies move into the spotlight because the margin separating the candidates is of the same size as the measurement inaccuracies. This difference was well within the range of uncertainty in the voting and counting procedures, and the outcome of this election was buried in the uncertainties of the electoral process. The voting and counting procedures in Florida, and in most other voting venues as well, do not have sufficient accuracy to identify a winner in the face of such a small difference in the vote totals. In the context of the voting and counting procedures in place in Florida, the election was a tie. This conclusion has since been upheld by the prolonged recounts that have been conducted by several newspapers. This media consortium determined that the outcome depends on how you count the ballots.[6] An examination of the contested ballots under more restrictive standards of interpretation yielded one winner, and under less restrictive standards another winner.

The inability of the normal voting and counting procedures to determine a winner of an election does not, of course, mean that no

---

[6] *New York Times*, 11 May and 12 November 2001.

one takes office. What it does mean is that the person who takes office is then selected by other means. Americans are assured that they will have a new president inaugurated on the constitutionally specified date. The backup procedures called into play when the vote is inconclusive include legislative and judicial interventions. Legislative solutions simply acknowledge the inconclusiveness of the election and substitute a vote of the legislative body, be it a state legislature or the US Congress. The courts may make judgments whether to include or exclude 'gray' votes, for example votes represented by an incompletely detached chad or those that had an unclear postmark.

In the case of the closely contested presidential election of the year 2000, the US Supreme Court effectively selected the new president simply by ordering that the recounting of votes in Florida must stop at a point when George W. Bush led by a few hundred votes. Uncertainty cannot stand indefinitely in the way of selecting a president, and the constitutionally prescribed procedures contain an explicit recognition that a decision must be made in the face of uncertainty. There is no option 'to study the problem' indefinitely into the future, although many would argue that there was ample time to carry out a manual inspection of the ballots passed over earlier by the machines that first counted the ballots.

The very close US presidential election of 2000 has raised the question of whether we can design a better measuring instrument, one that would yield a more complete and accurate count of the votes cast. The uncertainty of the close election has forced the voters and their governors and legislators to think about how the process of voting and the counting of the votes can be improved; the uncertainty has served as a spur to creativity and ingenuity. If all elections yielded winners who carried 60% of the vote, the present yardstick would be adequate. It was clearly inadequate in the 2000 presidential election where each candidate garnered 50%, give or take 0.0001% . The national experience with an inadequate measuring device has already led to many suggestions as to how the process of voting can be improved in America.

BEYOND MEASUREMENTS

The discussion of uncertainty in most science and statistics textbooks ends after addressing the uncertainties of measurement, calibration, and repeatability. While these concepts of uncertainty are important, they may not, in fact, be the greatest source of uncertainty in a scientific enterprise. The larger problems of uncertainty arise in other ways, quite beyond the precision and accuracy of measuring devices. One of these issues centers on how we aggregate one-at-a-time observations into a bigger picture. The pointillist style of painting, so well developed by the French artist Georges Seurat (1859–1891), is at close range just a collection of paint daubs. From a distance, these dots merge to become a representational picture and a work of art. The question reduces to "How can we say something important about the forest based on a collection of measurements made on individual trees?"

In early 1998, a newspaper headline declared "1997 The Warmest Year On Record".[7] Upon reading the article, one learned that an analysis of millions of temperature measurements made at weather stations around the world and on buoys and ships at sea indicated that the global average temperature for 1997 exceeded the average temperature for any other year since systematic temperature measurements have been made and collected. But what exactly is meant by an annual global average temperature? How can an average temperature be determined? Why is there uncertainty in such a number?

Let us start with taking the temperature at a single location at a given time. You can easily envision a person waking up in the morning and glancing through the window at the outdoor thermometer to get a clue as to how to dress for the day. If the person wrote that temperature down every day, at the end of the year all the readings could be averaged to determine the yearly average of the morning temperature. However, on occasions our diligent observer forgets to read the thermometer, or fails to put on glasses before taking the reading, or writes the reading

[7] New York Times, 9 January 1998.

down incorrectly, or goes on vacation, or accidentally breaks the thermometer. All of these sins of commission or omission lead to what are known as data gaps or data errors, intervals of time for which there are incorrect or no observations. In addition, if the thermometer needs to be replaced, perhaps our observer decides to buy a different style of thermometer that has a different sensitivity and calibration and thus gives a slightly different reading than the original thermometer would for the same temperature. And how should we compare a temperature taken at one house where the thermometer is in the shade with the temperature taken next door where the thermometer is in the sun? All these problems can be addressed and accommodated, but each adds a touch of uncertainty to the determination of an annual average temperature for that location.

Another consideration in determining an annual mean temperature at a site relates to the timing of the temperature measurements, and how temperatures taken at different times should be averaged. The observations by the person diligently taking temperature readings every morning as he picks up the newspaper, when averaged over an entire year, would give us a yearly average, but that would be the 6:00 a.m. average temperature, not a representative twenty-four hour average. Our observer might also note the temperature when he returns home at 6:00 p.m., and from both his morning and evening temperatures estimate a daily average. But he could do an even better job if in addition to the morning and evening observations he made other measurements at noon and midnight. You see the point: a good daily average temperature at a place must be determined from observations that represent all parts of the temperature swing between day and night. Once we have good estimates of average daily temperatures at a location, we can in principle easily construct monthly and yearly averages.

In the real world, of course, we do not rely on amateur observers to read their outside thermometers each day. That job has been assumed by professional meteorologists who work at official weather stations. At such stations, where taking the temperature is a principal activity, the temperature is recorded at least hourly and sometimes continuously, with a standard thermometer placed in a standard housing. Detailed hour-by-hour information is valuable in understanding the causes of climate change over decades and centuries, because it is important to be able to determine whether a warming results from ordinary days but warmer nights, warmer days but ordinary nights, or a warming that has affected both daytime and nighttime temperatures. Each pattern arises as the response to a different climatological stimulus. Just as with our hypothetical amateur observer, weather stations also experience occasional broken thermometers, short-term down time caused by power failures, closures because of inadequate funding or war, and relocation from urban to suburban sites, all of which creates data gaps and errors and contributes to uncertainty about the measurements.

Another source of uncertainty in establishing an annual global average temperature relates to where the temperature was measured. Temperatures are not taken at every point on the surface of the Earth but rather are collected from only a few thousand sites irregularly distributed around the globe. There are many more weather stations in Europe or the USA, for example, than in Siberia, the Amazon River basin, the Sahara desert, or Antarctica, all territories of comparable or larger area than either Europe or the USA. Let us assume for argument's sake that there are 400 weather stations in Europe, and 100 in Siberia, the Amazon, Antarctica, and the Sahara together. If we take a simple straightforward average of the readings of all these stations, we would have a result heavily weighted by the large number of stations in Europe compared with elsewhere, and the average would be regionally biased by the larger number of measurements in Europe. Clearly, one must be careful not to let abundant observations from one region of the Earth dominate the calculation of an average for the entire Earth. One must let the fewer measurements from the remote and inaccessible areas carry more weight in the average than do the many measurements that come from the more easily observed areas of the globe. A better approach would be to divide the Earth's surface up into regions of equal area, determine an average temperature for each area, then determine a global average from the average temperature of each of the equal area regions. But it is clear that if the average in one of the regions is determined from only a handful of weather stations and in another by hundreds, there will be more uncertainty about the regional average determined from the few observations than for the average determined from the many. The global value determined from the regional averages will reflect the variable uncertainty associated with the regional averages.

The issues associated with the geographic distribution of observation sites have made it impossible to determine a global average temperature directly from thermometer readings prior to about 1860 for the simple reason that there were not enough weather stations established in the southern hemisphere to be able to represent that large region of the Earth adequately in a global estimate. Since 1860 or so, enough stations with adequate geographic distribution have been operational to enable the determination of a representative average temperature for each location and, through aggregation, for the Earth as a whole. Each carries some uncertainty of its own and adds a touch of uncertainty to the global determination.

COUNTING PEOPLE
Other examples of large-scale data collection and aggregation help us to visualize the complexities and uncertainties inherent in any assembly and analysis of large numbers of observations. For example, the question "How many people live in the USA?" comes up every decade when the constitutionally mandated national census is undertaken. Like the temperature, the population is a slippery target, never standing still, changing all the time.

Census taking, like temperature taking, has many sources of uncertainty. A central strategy of census taking is to determine the number of occupants of each residence in the nation. Therefore, via the mail, a census questionnaire is sent to every residential address asking the essential demographic questions. However, some recipients will deposit the form directly in the circular file, and others will forget to fill it out. Some respondents will not answer truthfully, for fear that they are violating some regulation about allowable occupancy levels, or because they know that some of the residents are illegal immigrants. Some people will live at addresses not identified as residential, while others, such as the homeless, will have no address at all.

Undercounting is not the only source of error and uncertainty in the census. Overcounting also occurs. Children of divorced parents sometimes get counted on each parent's form; students away at boarding school or university sometimes will be counted both at home and school. The possibility of multiple filings from the same household exists. In the 2000 census, forms were mailed to addresses compiled from a number of different lists. In order to be complete as possible, the Census Bureau merged its primary mailing list with information from the US Postal Service and local governments. Although efforts were made to exclude duplicate mailings, some undoubtedly slipped through. Forms were also available at government offices and at convenience stores. It was also possible to respond to the census via telephone or the Internet; thus the opportunities for duplicate filings were abundant.

Follow-up inquiries by telephone or by census-takers walking the neighborhood will improve the data collection somewhat, but in the final analysis there will remain a census product that is incomplete and inaccurate and, therefore, uncertain. Just as the direct enumeration of votes does not always lead to a clear election result, so too will the census strategy of direct enumeration of each and every person in the country always yield an imperfect product.

The final 2000 census count for the USA yielded some 281 million people, after deleting some 3.6 million overcounts caused by duplicate or multiple filings. The Census Bureau estimated that the net undercount probably was in the range of 2.7 to 4 million people, or about 1 to 1.4% of the people who were tallied. And just as in the election vote counting, the imperfections in the census were not uniformly distributed among all segments of the population. The uncertainty in the final census estimate derived unevenly from different regions and populations. The undercount rate for Hispanics was estimated to be in the range 2.0 to 3.5%, for African-Americans about 1.6 to 2.7%, for American-Indians 2.8 to 6.7%, and for children under 18 about 1.2 to 1.8%.[8] The next chapter examines how statistical techniques might be used to improve the census product, in the sense of making it more accurate. But potential improvements also have political implications that can deter the pursuit of a more accurate census.

READING BETWEEN THE LINES

Let us turn to a practical problem that arises in working with aggregations of observations. Consider, for example, the problem of determining, after the occurrence of an earthquake how wide an area experienced ground accelerations that exceeded a certain level. Ground vibrations are typically greatest near the epicenter of an earthquake and get smaller with distance away from the epicenter. How big is the region around the quake that experienced big accelerations? This determination is very important, because it plays a role in establishing construction standards for schools, hospitals, and nuclear power stations in seismically hazardous areas. The measuring device relevant to this undertaking is an instrument called an accelerometer. These instruments are deployed widely in earthquake-prone regions and may number in the hundreds in cities such as Los Angeles that are situated in areas of high seismic hazard. Away from such locations, these instruments are also deployed, but in far fewer numbers. There are probably more accelerometers in Los Angeles than in the entire state of North Dakota.

These instruments are not all from the same manufacturer, nor even of the same design, but through calibration seismologists can assume they will adequately determine the acceleration (within an acceptable uncertainty) felt locally near an earthquake, or remotely many hundreds of kilometers away. So after an earthquake,

[8] *New York Times*, 12 January 2001 and 14 February 2001.

seismologists gather all the readings together and register them on maps at the locations of the instruments. Then comes the hard part: from this collection of scattered readings at many distances and directions from the quake, how do the scientists estimate the area that experienced accelerations in excess of a certain value?

Let us take an acceleration of 0.1 g (one tenth of the acceleration of gravity) as the value of interest. At one accelerometer site we note a reading of 0.12 g, definitely in excess of the level we have selected, and at another site more distant from the quake another station registered a reading of 0.03 g, clearly lower than the selected value. In fact, in the entire array of instruments some registering above and others below 0.1 g, there may not be a single instrument that registered exactly 0.1 g. The task the seismologists face is to draw a line through this irregular array of points that separates values that exceed 0.1 g from those that are less. In effect, the line is the seismologists' best guess as to where an instrument would have recorded an acceleration of 0.1 g. The area enclosed by that line can be easily determined once the line has been drawn, but considerable judgment needs to be exercised in placing the line properly between the locations with instrumental readings.

The process of estimating where a given value will occur between two known values is called *interpolation* and is similar to reading a thermometer when the top of the liquid in the glass tube falls between the markings on the glass. In the case of the thermometer, we can usually assume that the temperature increment beyond the lower marking is directly proportional to the fraction of the distance between the two markings that bracket the top of the liquid column. For an earthquake, one might think that the lines of diminishing acceleration would just be circles at increasing distances from the epicenter, something like the ripples spreading out when a stone is tossed into a pond. But the pattern of vibrations from earthquakes is never that geometrically simple. Ground accelerations are very sensitive to the type of soil or rock at the surface, and whether it is saturated with water. The vibration levels also depend on the geological structure of the subsurface between the earthquake and the recording instrument. Many tens of kilometers may separate two instrumental readings, and all of these factors may vary considerably between the two sites.

This uncertainty in determining the area that experienced a certain level of ground accelerations has further consequences. Let us, for the sake of argument, say that the earthquake that caused such accelerations was determined to be a magnitude 6 event, and that earlier studies had determined the lesser areas experiencing this same level of acceleration from magnitudes 5 and 4 events, each with some uncertainty for the same reasons as mentioned above. Using the results from all of these investigations, which showed a larger area feeling such accelerations as the size of the earthquake increased, we want now to estimate how large an area would be similarly shaken by a really big earthquake, say a magnitude 7 or 8 earthquake. It is apparent that uncertainties in the area determinations from the three smaller earthquakes will introduce some uncertainty in the estimate of the shaken area for an earthquake bigger than any experienced so far, an earthquake that is yet to come.

In designing a structure, such as a nuclear power station, to withstand earthquake shaking, it is important to establish a careful estimate of the vibration levels that the structure may experience. This, in turn, requires that both the location and the size of the largest earthquake that is likely to affect the construction site be determined. This event, known as the 'maximum credible earthquake' that the region is likely to experience, is by definition larger than any earthquake the region has already experienced. Estimation of the maximum credible earthquake requires a speculation outside the realm of what we have observed so far, beyond the range of our experience to date. This is a process called *extrapolation*. The estimate of the maximum credible earthquake carries a lot of dollar signs with it; a larger estimate means a sturdier and more expensive building, a smaller estimate a less expensive one. From this perspective, the uncertainties associated with the design and calibration of the measuring device, the accelerometer, are the least of the relevant worries. Those uncertainties are overshadowed first by the uncertainty in determining the area of critical shaking from the readings of accelerometers deployed in the field and then by the uncertainty in extending those observations to predict how large an area would be affected by a larger but yet-to-be-experienced earthquake. All of those uncertainties are, in turn, compounded by the additional uncertainty in estimating the size of the maximum credible earthquake that the region might experience. Clearly, the uncertainty associated with the precision and accuracy of the accelerometer readings are small when placed in the context of the overall uncertainty in the important quantities derived from those readings.

All of these uncertainties, however, do not preclude decision-making. Public officials determine where and how to build new schools, freeway overpasses, and power plants. The conventional engineering wisdom of the day, along with economic, demographic, and political factors, guide construction standards, with a healthy cushion of safety added on to allow for the uncertainties. As additional experience accumulates, the lessons learned from successful and, yes, failed designs lead to revision of the construction standards. We learn from our mistakes and from our successes. The standards, indeed, are not set in concrete.

————————————

So now we have made our way through the intricacies of measurement. We have considered the suitability of a particular measuring tool for the measurements we undertake, and ways in which many individual measurements can be aggregated to tell bigger and more significant stories. The next chapter addresses the quantitative descriptions of ensembles of measurements using the tools of statistics and probabilities.

6   A fifty–fifty chance

> This country is hungry for information; everything of a statistical
> character, or even a statistical appearance, is taken up with an eagerness
> that is almost pathetic; the community have not yet learned to be half
> skeptical and critical enough in respect to such statements.
>
> Francis A. Walker, Superintendent of the 1870 US Census

The previous chapter focused on measurements. Here we will talk
more about how to extract quantitative information from a collection
of measurements. The discussion will lead us into topics as diverse as
flood and earthquake frequencies, election polls, and census taking.
What can a poll of a small number of registered voters tell us about the
likely outcome of a forthcoming election? What can we learn from the
past history of flooding along a river that will give some indication
of what we might expect in the future? Uncertainties are associated
with each topic, uncertainties that arise from different sources and
are quantified in different ways.

There are many processes and pathways that lead to ensembles
of measurements. One very common source is simply making a number of measurements on a single object – each student in a class at the
local elementary school measures the height of their teacher, or all
seismograph stations in a region estimate the magnitude of yesterday's
earthquake. A second common ensemble comprises one-time measurements of a number of different objects – perhaps the weight of
each student in the class on the first morning of the new school term,
or the concentration of arsenic in each water well in the county on
a given day. A third type of data collection is a set of observations
taken over time – the average weight of eight-year-old students at the
beginning of each school year over the long history of the school, or
the daily water level of the local river over the past seventy-five years.
How can we characterize these ensembles quantitatively? What can
we learn from each of these collections of measurements?

STATISTICS AND PROBABILITIES

Statistics are quantitative descriptions of a data ensemble, calculated by defined mathematical procedures. When the students pooled their individual measurements of the height of the teacher, the measurements clustered around 5 feet 9 inches (1.75 meters), some a little less, some a little more. The descriptive statistic that is useful here is simply the average of all the measurements: add them up and divide by the number of measurements. The average is then the class's best estimate of the height of the teacher, and the spread of the observations on either side of the average is a measure of the uncertainty in the determination. The spread is an uncertainty that arises from the variable skill of each student observer in using and reading the measuring device.

The second ensemble, the weights of the students in the class, showed a range between fifty-three and eighty-one pounds, and the average weight was determined to be sixty-six pounds (thirty kilograms). In this case, the range exists principally from the variability of weight between the class members, although some small part may have been due to reading the scale. We can learn more from these measurements than just the average value. We can also determine how much each student's weight departs from the average, and we can calculate an average departure. This gives a feeling for how tightly clustered or how widely distributed the individual student weights are when compared with the average. Do we have a monotonous class where everyone weighs almost the same, or do we have a highly variable class of overweight students and anorexics?

We can also determine the range of weights around the average that contains the middle two-thirds of the students, or find the weight for which half the class is heavier and half lighter. All of these are quantitative descriptions of the ensemble of weight measurements. Such measurements and their statistical descriptions can be used as a starting point for scientific investigations of this group, and of eight-year-old children elsewhere. How representative are the students at this school with those at other elementary schools in the county?

Can we correlate the average weight, or the range of the weights, with socioeconomic factors across the county?

Probabilities are estimates of likelihood. In the context of our eight-year-old children, we might want to know the likelihood that the weight of a student will be between sixty and seventy pounds (twenty-seven and thirty-two kilograms). An analysis may reveal that there is a 75% probability that a student's weight will fall within that range. But the same calculation also says there is a 25% probability that it may not. There is some uncertainty, in this case one chance in four, that the weight of a randomly selected student will fall outside of that range. Probabilities provide a tool with which we can quantify uncertainty.

Many scientists use the language in the table to describe quantitative estimates of probabilities in terms of the associated uncertainty. Uncertainty on occasion is black or white, but usually it comes in shades of gray. Or as Bertrand Russell said, "When one admits that nothing is certain, one must also add that some things are more certain than others." Using the table as a guide, the probability that one might win the lottery could be called 'extremely unlikely'; the probability that the sun will rise tomorrow is a 'virtual certainty'. The probability that your children will go to college falls into the gray terrain between the extremes.

Language of probability

Probability (%)	Terminology
< 1	Extremely unlikely
1–10	Little chance or very unlikely
10–33	Some chance or unlikely
33–66	Medium likelihood
66–90	Likely or probable
90–99	Very likely or very probable
> 99	Virtual certainty

The Intergovernmental Panel on Climate Change (IPCC) is an organization created some two decades ago by the United Nations and the World Meteorological Organization. The IPCC was established to assess the state of knowledge about climate and the factors that affect it, to estimate the range of consequences of climate change both globally and regionally, and to provide a range of scenarios about the future that might be the outcome of certain demographic and economic pathways through the twenty-first century. In the assessment of various aspects of global climate, the IPCC scientists have taken special care to quantify the uncertainty associated with each set of observations and each projection for the future. The observation that atmospheric abundances of carbon dioxide have increased throughout the twentieth century is ranked as a 'virtual certainty'; decreases in soil moisture as a consequence of warmer summers in the northern mid-latitudes is assessed as a 'probable' projection. Each component, each process in a complex system such as the global climate system has a different degree of uncertainty, some parts very well known, and others less so.

Our desire to assess probabilities in everyday life is deeply embedded. What are the chances of rain tomorrow? What are the odds that the winning golfer in the next British Open will score under 270? What is the likelihood of surviving a surgical procedure? Much of the quantification of uncertainty for future events comes from an analysis of the statistics of such events in the past. An analysis of winning scores in all previous British Open golf tournaments would enable an easy first assessment of the odds for a sub-270 winner. And when a physician tells a patient "You have a 50% probability of surviving the operation", the doctor is probably summarizing the history of other patients who have undergone that operation. On the one hand, if the operation has been performed thousands of times, and half the patients died during surgery, the estimate of 50% survival probability is a robust estimate derived from considerable experience. If, on the other hand, the operation has only been performed twice, and one patient died, the probability derived from that past experience is still

50%. However, the *confidence* one has as to whether that estimate of the rate of mortality is indicative of the outcomes of the next 100 operations is very low.

The old adage "we learn by experience" is usually true, and statistics help us to measure how much we have learned. If an operation has been performed thousands of times over several decades, and the mortality rate was initially high but over the years has fallen dramatically, the overall mortality rate is probably not a fair indicator of what the next patient will face. With experience, we learn more about how things work. We hope the doctors who perform operations improve their skills, acquire better tools, and devise ways to avoid certain perils, so that as experience accumulates, the survival probability will also change.

We can look at the daily weather forecast as another example of a procedure that has improved with time. When we watch the weather report on television and hear that there is a 90% probability that it will snow tomorrow, we plan accordingly, confident that we will see some of the white stuff on the ground. But it was not always so. Decades ago, the weather forecast was a common target of comedians. To be fair, forecasting has moved well beyond pure statistics, such as a forecast of rain for tomorrow based on the observation that over the past 128 years it had rained on that date 65% of the years. Today we have much more confidence in forecasts because we have satellite cameras giving synoptic views of continent-wide weather systems, a global network of weather stations sending measurements of temperature and air pressure to large computers that churn out reliable predictions hour by hour. In short the probability has improved because we have better instruments that provide more information on a timely basis to ever-improving computer models that can process it to yield a forecast that has a high probability of being on the mark.

In the examples of the surgical procedure and the weather forecast, the probabilities for success have improved over the years. However, this fact surely would never have been an excuse to postpone surgery or not to make a weather forecast. We clearly recognize that, while future knowledge will likely serve us better, we usually do not have the luxury of being able to wait for future knowledge. Incomplete knowledge is usually better than none.

MISUNDERSTANDING PROBABILITIES

Unfortunately, probabilities and the basis on which they are calculated are frequently misunderstood. I recall an old joke about a patient facing risky surgery. His surgeon explained that there was only a fifty–fifty chance that he would survive the operation. The patient mulled over this information and then told the doctor that he would proceed, but wanted to be scheduled immediately after someone else with the same infirmity had died in surgery. His logic? If there was one chance in two that a patient would survive, he wanted to be the one to balance off the previous death. The flaw in this logic can be easily seen: his survival (or death) in surgery has absolutely no link to the fate of any previous patient. Nor will the fate of any subsequent patient be dependent on the outcome of his surgery. In much of probability theory, every event is considered an independent event, and not conditional on another event. In a sequence of coin flips, the outcome of the next flip is independent of the number of heads and tails that have already occurred. This independence of event outcomes is a commonly misunderstood or unappreciated concept. Probabilities are also prone to misunderstandings and misinterpretation, particularly when cast in unfamiliar quantitative language. As an illustration, let us look at the hemoccult test that in simple terms looks for blood in the feces as a possible indicator of cancer of the colon and/or rectum (colorectal cancer).[1] The statistics of occurrence of colorectal cancer show that its frequency in the general US population is about 300 cases in every 100,000 people, and the hemoccult test will show positive (i.e. show blood in the feces) in about half of those persons with colorectal cancer. If a person does not have this disease,

---

[1] This example follows closely the presentation in the following paper: Hoffrage, U., Lindsey, S., Hertwig, R., and Gigerenzer, G., Communicating Statistical Information. *Science* vol. 290, pp. 2261–2262, 2000.

there is still a 3% chance that he or she will test positive, so there is also the possibility that one can show positive in the hemoccult test even if free of the disease. This simply indicates that there are other reasons besides colorectal cancer for the presence of blood in feces.

A physician, staring at a positive hemoccult test, must determine the likelihood that the patient actually has colorectal cancer. The problem is not as difficult as it sounds. Of the 300 who actually have colorectal cancer, about half, or 150, will test positive. Of the 99,700 who do not have the cancer, 3% or about 3,000 will also test positive. The total of the positive tests is 3150, of whom 150 actually have cancer. Thus the likelihood that a person testing positive actually has colorectal cancer is 150/3150, or about one chance in twenty. But knowing that half of the people who do have the cancer will test positive, the hemoccult test is a flag to physicians to follow up a positive test with other types of test, such as a colonoscopy, to determine whether the positive hemoccult result was caused by the presence of cancer or was merely one of the nineteen chances out of twenty where a positive test outcome is not an indication of the disease.

We would probably not be very surprised if the average patient found this excursion into medical probabilities a little baffling. However, it may be more than disconcerting to learn that a third or more of physicians, when given the above facts and asked to estimate the probability that a positive hemoccult test indicates colorectal cancer, failed to reach the one in twenty result. The language of probability (although not necessarily the concepts) is a barrier to understanding, and misunderstanding always contributes to uncertainty.

THE HUNDRED-YEAR FLOOD

Let us consider another example. When a news reporter remarks "That was the second occurrence of the 'hundred-year flood' in less than a decade", many a listener wonders how can that be? How can a second 'hundred-year flood' occur less than a century after the first? The terminology 'an XX-year flood' really signifies that, *on average*, one will wait XX years before experiencing a flood of that same magnitude. It also says that, *on average,* there will be only one flood of that magnitude in XX years. When a second flood occurs sooner than XX years, it has come earlier than the average return interval, and of course an interval between such floods greater than XX years is simply one that was greater than average. The descriptive terminology should not be interpreted as ruling out the possibility that more than one will occur, nor that it guarantees that one will occur in the designated interval. The terminology shines a spotlight on the average behavior of a watershed.

Hydrologists have been slow to cast these concepts in terms of likelihoods rather than return periods. The hundred-year flood can also be described as the flood that has a one in a hundred chance of occurring in a given year. This description helps the non-technical consumer of such information to understand why more than a single hundred-year flood can occur in a century. In any given year there is a 1% probability that such a flood might occur. No matter when it occurs, it is an improbable event.

The probability estimates set out for a given watershed are based in part on the historical observations of how high floodwaters rise on a year-to-year basis. In most years, the waters rise and fall seasonally but stay within the banks of the stream or river channel. On occasion, they spill over the banks and inundate some of the surrounding area. On rare occasions, they cover the entire valley or flood plain and rise up onto the valley flanks. For the more frequently occurring water level fluctuations, say a level that is reached in three out of every four years, we have some confidence, based on a century or more of observations, in stating that there is a 75% probability that waters will rise to that level in a given year. For the occasional spillover that has occurred twelve times in the past century, we again estimate with some lesser confidence a 12% probability that a spillover will occur in a given year.

But what about the probabilities for another event comparable to the substantial floods that inundated the hypothetical village of

Podunk Crossing in 1924 and again in 1976? We can mention that there were two events in the century, or that such floods were separated by fifty-two years, but we surely must feel insecure in stating that this single measurement of a recurrence interval is a robust estimate of the *average* recurrence interval. And what about the really catastrophic flood that rises well up on the valley walls and would inundate houses up to their second story windows? Podunk Crossing, first settled in 1887, has never experienced such a flood in its entire 127-year history. But could it fall victim to such an event? And if so with what frequency? The answers to such questions are derived from a probability distribution estimate based on the frequency of occurrence of smaller floods, and some theoretical ideas about the relationship of frequency of occurrence versus flood size. Such an analysis for the Podunk Creek watershed might indicate that such a flood corresponds to a 'two hundred year flood', signifying that, on average, such a flood might occur every two hundred years, or that in a given year there is a 1 in 200 probability that such a flood will occur.

Similar analyses take place in estimating the probability that an earthquake of a certain magnitude will occur. Although seismologists do not speak of the 'hundred-year earthquake', the concept is the same. A calculation of the maximum likely ground acceleration that a soon-to-be-built school or nuclear power station might experience involves an estimate of the maximum credible earthquake that might strike the location in the lifetime of the structure. This estimate is nearly always based on the frequency of occurrence of smaller earthquakes of various magnitudes that the region has experienced, extrapolated to the higher credible magnitude, which the region has not yet experienced. If the lifetime of the envisioned structure is estimated at fifty years, it would be unwise to plan only for the 'fifty-year earthquake' because there is a non-trivial probability that a 'hundred-year earthquake' might occur earlier than its average recurrence interval of one hundred years.

One of the principal assumptions of such calculations is that the underlying physical processes that govern flood and earthquake occurrence remain unchanged over long periods of time. In the case of earthquakes that may be a safe assumption because the phenomenon involved, the slow shifting of Earth's tectonic plates over millions of years, is governed by processes deep in Earth's interior that undergo change on geological time scales. It is naive, however, to think that flooding in a watershed remains a stationary process, a technical term signifying that the conditions that lead to flooding remained the same over the time interval for which observations about floodwaters have been gathered.

Many regional characteristics that affect flooding undergo change as a region is settled and developed. A factor in the flooding equation is how much of the water in a rainstorm or from melting snow infiltrates into the subsurface, and how much runs off over the surface, eventually making its way into the principal waterways of the region. The telltale signs of development, the clearing of woods for agriculture or the paving of fields for streets and parking lots, all diminish the infiltration and increase the runoff. Land-use changes far upstream can contribute to an increased downstream flooding potential. So even if the local weather remains unchanged over the decades, with urbanization the probability of flooding is slowly increasing.

The possibility that the mean annual precipitation is also slowly changing through perturbations of the global climate system adds yet another factor to lend uncertainty to the flood probabilities. Another factor in the infiltration/runoff partition is the duration of rainfall events. A slow and steady rainfall leads to greater infiltration, whereas a short downpour leads to greater runoff. One apparent manifestation of climate change that has been well documented[2] is the increasing frequency of severe rainfall events. These events are defined as ones in which two inches (five centimeters) or more of rainfall occurs in a twenty-four hour interval. The fraction of the area of the USA that experiences such an event in a given year has been creeping upward for more than ninety years.

---

[2] Karl, T. R., Knight, R. W., and Plummer, N., Trends in high-frequency climate variability in the twentieth century. *Nature* vol. 377, pp. 217–220, 1995.

All of these factors tell us that the flood probabilities, just as with the survival probabilities associated with surgery, are not standing still. We must try not only to establish the probabilities of flooding based on historical occurrences but also to recognize that such probabilities are a moving target that are changing as we alter conditions locally through land development and globally through anthropogenic contributions to climate change. People who build houses on floodplains, zoning codes that regulate such construction, and insurance companies that offer flood insurance must recognize the changing pattern of risk through time.

ESTIMATING FROM SAMPLES

An undertaking now ubiquitous in elections in many countries is the polling carried out in advance of elections, to determine how various candidates are faring. The question asked is familiar to anyone who has participated in such polls: "If the election for the Representative of the seventh district were held today, would you vote for Mary Jones or Alice Smith?" Shortly thereafter, we hear on the evening news that "a poll of 600 likely voters indicates that 42% favor Mary Jones, 44% favor Alice Smith, and 14% remain undecided. The margin of error in these numbers is plus or minus 4%, so the two candidates are in a statistical dead heat."

What do these numbers mean? The uncertainty in the estimate of each candidates' current polling strength is indicated by the margin of error. The margin of error arises because in this 'sample election' the votes of 600 people were tallied, whereas in the real election, some weeks away, more than 200,000 people will probably vote. The results of the sample election could also have been reported in the following language. "A poll of 600 likely voters indicated that, if the election were held today, there is a 95% probability that Mary Jones' vote total will fall in the range 38–46%, and Alice Smith's in the range 40–48%. Because there is so much overlap in the estimated ranges, there is real uncertainty about the eventual outcome; consequently, the election, were it to be held today, would be too close to call with confidence."

The statement "there is a 95% probability that Candidate X's vote total will fall in the range 38–46%" tells us that if the pollster repeated his experiment 100 times, each time selecting and calling 600 likely voters to determine whom they would vote for, in 95 of those repeats the number of respondents choosing Candidate X would fall in the range 38–46%. In only five repeats would the result fall outside that range. In other words, there is a 5% probability that the poll results do not accurately reflect Candidate X's standing. If one wanted to have less uncertainty, say to raise the probability from 95% to 99% that another poll would fall within this specified range, more than twice as many people would have to be polled. Clearly there is a cost to reducing the uncertainty in polling. Without increasing the number of people polled, one can also estimate a percentage range in which the pollster was 99% confident that Candidate X's vote would fall, but that percentage range would, of course, be bigger than that for 95% confidence. There is clearly a tradeoff between the range that will bracket Candidate X's likely share of the vote and the probability that the tally will fall in that range. The bigger the range, the higher the probability. Conversely, if the range is fairly narrow, the probability that the vote will fall within that range is correspondingly lower. An easy analogy is to think of throwing darts at a dartboard. The probability of hitting anywhere on the board, a big area, is of course much greater than hitting the bulls-eye, a small area.

Most pollsters find that the tradeoff between the number of people polled and the probability that the polling result is accurate comes at the 95% probability level. A pollster will have 95% confidence that a sample of around 600 voters will yield an estimate of a candidate's standing within a range of uncertainty of approximately plus or minus 4%. Polling fewer voters would be cheaper but have a greater margin of error. A larger sample would lessen the range of uncertainty but would be more costly. Moreover, the improvement that accompanies a larger sample comes very slowly. To cut the range of uncertainty in half, one must poll four times as many people. And one must remember that polls, in addition to not being prohibitively expensive, must also be timely. Seeking opinions of an ever-increasing number of people would fail on both accounts and would only improve the margin of error slightly.

But enough of the arithmetic. Is the formal calculation of the range of uncertainty, the margin of error as reported by the pollster, really a good estimate of the uncertainty about where the candidates stand at the time of the poll? Not necessarily. The answer depends critically on the skill of the pollster in selecting whom to poll. Whenever one tries to estimate a characteristic of a large group by measuring that characteristic on a small sample drawn from the large group, the credibility of the estimate rides on how representative of the large group the small group actually is. In short, careful sample selection is at the heart of reliable estimation.

A random sample of names taken from the telephone book will include many sources of error: about half of such people do not even vote, some because they are underage, some because they are not registered, some because they are not citizens, some because they have moved away, some because they feel politics and elections are irrelevant to their daily lives. Other aspects of good sample selection are demographic. The sample should be geographically, economically, politically and culturally representative. If County A has twice as many eligible voters as County B, but the average voter turnout in County A and County B has been about 47% and 55%, respectively, then this information should guide the selection of the polling sample. Pollsters know that the quality of their poll results can be improved by careful sample selection or by polling a larger number of people, or both. Improvements in sampling cost money, and there is the inevitable tradeoff between greater accuracy in the poll and the costs incurred to achieve it.

There always seems to be some incredulity that a well-selected sample of a few thousand people can successfully predict the outcome of a national election in which one hundred million votes are cast. Probability theory is very clear on this point, however. The fraction of the population sampled is not the important issue; it is the actual size of the sample, the number of people who have been interrogated, that determines the probability that an estimate determined from the sample will be representative of the overall population. The characteristics of a very large group can be estimated with surprising accuracy, say within a margin of error of a few percent, by examining a sample of a few thousand, no matter whether that sample has been drawn from a population of one hundred thousand or one hundred million.

I once tested this proposition with my son, who at the time had a great interest in coin collecting. The question we posed was the following. Could we estimate the relative number of pennies minted in each of ten recent years from the pile of pennies that resided in the family cookie jar? The mint production of pennies varies from year to year, depending on the need for pennies in circulation, but they usually run in the hundreds of millions. Can a cookie jar full of a few thousand pennies serve as a sufficient and representative sample of the hundreds of millions of coins produced each year? Yes indeed. The cookie jar collection was more than adequate to estimate the production variability from year to year, within a very small margin of error. Probability theory helps us to decide how big a sample is necessary to achieve a result within a certain margin of error, and a cookie jar full of pennies went a long way toward producing a very accurate result.

Polls that track election sentiment over time reflect the changing attitudes as the campaigns unfold. Earlier opinions are sometimes altered, and the 'undecideds' eventually make up their minds. Election campaigns, like most aspects of life, are never static. Factors that may lead to ups and downs in the poll standings include the money available to candidates to deploy their message, mistakes and misstatements by the candidates, and the treatment of the candidates by the media. The eventual outcome is a moving target, and candidates, in particular, pay close attention to changes in tracking polls and the interpretation of what is leading to the changing sentiment. Decision-making by voters is a most flexible process, and seldom do campaigns end with the same poll standing that they began with.

CENSUS TAKING

In the previous chapter, I introduced the decadal census in the USA to illustrate the uncertainties in counting. That there ultimately is uncertainty in the actual number of people living in the USA at the end of the year 2000 should come as no surprise to anyone, quite apart from the fact that births and deaths during the counting period make 'a perfect count' elusive. Counting people in a census, just as counting votes in an election, runs into limits on the accuracy of the final tally. However, because the geographic distribution of the population is a central factor in the drawing of legislative and congressional districts, and therefore in the relative representation of different demographic groups in those important state and national law-making bodies, a non-trivial shift in political power derives from an incomplete and imperfect census.

A debate about how to improve the national census occurs each decade at the time of the constitutionally mandated count. One view is that we must basically stick with the attempt actually to enumerate each person through improved, more intensive direct interactions (mail, phone, Internet, on-site interviews), whereas the opposing view favors a sampling strategy to estimate the numbers of persons missed in the direct enumeration. For example, if mail questionnaires and census-takers were able to reach only 88% of the households in a neighborhood, an estimate for the remaining 12% of the households would be made from the characteristic responses of the 88% that had been enumerated by the traditional direct methods. The Census Bureau believes it can reliably estimate both overcounting and undercounting using survey data from only 314,000 households nationwide. These households comprise a sample of the many millions of households across the USA that the Census Bureau has designated the ACE group, an acronym that signifies accuracy and coverage evaluation.

Sampling techniques have been extensively employed to estimate much more about a neighborhood than just its population. Product marketing benefits greatly from being able to estimate household annual income, number of cars owned, the value of the average house, and similar economic characteristics from the official census data. Those who sell to the consumers of America have great confidence in the ability of incomplete census data to estimate the demographic and economic status of the complete neighborhood.

Again, the science behind sampling is relatively well understood and non-controversial. There is little doubt that a more accurate census could be achieved using both direct counting of the easily reached and sampling methods to estimate the over- and undercounts. The use of statistical methods to fill in the gaps in counting and coverage has, however, met considerable political resistance. This resistance arises not because the politicians who oppose it do not trust sampling techniques. Indeed, in pre-election polls, candidates for political office regularly and confidently employ sampling methods as current indicators of voter sentiment. Opposition to using sampling techniques in the census arises because there is political advantage to be had in an imperfect census achieved through direct counting alone. Maintaining a non-uniformity in the completeness and accuracy of the census across geographic settings and ethnic groups is an unstated but specific political goal of those advantaged by it. Those that are disadvantaged by it commonly include racial minorities, immigrants, and the poor, who are more prone to undercounting than other socioeconomic groups. Undercounting of these groups diminishes their representation in state legislatures and the US Congress.

In this chapter, we have had a glimpse of how measurements and observations can be described quantitatively through statistics, and how the quality and reliability of the information contained in ensembles of measurements can be estimated. This is a rich topic, one that has attracted literally thousands of essays, books, and monographs. Clearly we have opened the door to this vast intellectual arena only a crack. But even that narrow glimpse is sufficient for one to realize that certainty is generally beyond the reach not only of science but also every other endeavor based on the quantitative analysis of limited observations. Finite ensembles of observations usually define a range of interpretations and estimate the probability that the truth will lie within this range. A bigger range of possibilities, a bigger target so to speak, will have a greater probability of truth residing in it but may be of little use because of its lack of specificity. A narrower range of interpretations can be delineated, but the probability of finding truth is accordingly lower. These lessons about how much information can be gleaned from incomplete data provide a useful backdrop to the next chapters that address how we conceptualize and model complex systems.

7 How does this work?

Perplexity is the beginning of knowledge.

Kahlil Gibran

The quantification of measurements through statistical analysis, the discovery of a trend in temperature over time, or observation of a pattern in water pollution data displayed on a map – all motivate scientists to begin thinking and formulating ideas about what process lies behind the relationships they are observing. These ideas are initially simple and rudimentary, and they may lead to later testing through experimentation. In the next two chapters, we will explore the world of conceptualization and experimentation and gain insight into how uncertainty promotes creativity.

Scientists, indeed everyone, always operate with simplified concepts of the way things work. We call these simplified representations 'models', and they come in many forms: conceptual, physical, numerical. We receive imperfect guidance in model building from the real world, through incomplete, sometimes inaccurate, and occasionally conflicting measurements or observations about the phenomenon or system we are trying to understand. There is a continuous interaction between models and observation, with each undergoing adjustment in the face of the other. New observations lead to revision of a concept, and a new concept, in turn, suggests new experiments or observations to be made that will again put the concept to a test. It is this iterative back-and-forth interplay that generally improves understanding of a system, and which under some circumstances can reduce the uncertainty associated with system behavior. But when this fluidity is lost, for example when a scientist promotes or adheres to a concept in the face of considerable evidence to the contrary, or places too great a reliance on observations that are inaccurate or irrelevant, then progress stalls.

Most of the natural world is incredibly complex in its structure and organization. Contemplate for a moment the intricacies of a forest ecosystem, in which trees, fungi, microorganisms, birds, small rodents, large mammals, insects, ferns, snails, frogs, snakes, and many more types of life all coexist within the confines of the forest. Not only do they coexist, but they have interdependencies, where each provides some ingredient that allows another to flourish. Bacteria reside in soil, influence its chemistry and promote its development. Microbes also reside in the digestive tracts of larger animals and assist their digestion and metabolism. And all things great and small are influenced by the weather and climate of the region. Because of the complexity, it is extremely difficult for even the most capable ecologists to study a forest ecosystem in its full detail, and so they develop simplified concepts about the workings of the ecosystem, focusing on a few components and their interactions that are thought to be particularly significant. This conceptualization of the ecosystem web of interactions is called a model. To be sure, different ecologists may perceive the interactions differently, weigh the participation of the different components differently and, therefore, develop different models. Because of the complexity, the ecosystem is imperfectly understood and uncertainty about how it all hangs together is attendant.

The economy of a large industrial nation is likewise an intricate web of interactions between manufacturers, transport systems, wholesalers, retailers, banks, security and commodity markets, customers, farmers, labor, governments, tax codes, and much, much more. As with the forest ecosystem, it is virtually impossible to address the economy in its infinite detail, and so simplifications and aggregations take place. Imports and exports are lumped together as 'the balance of trade'; manufacturing capacity, wage levels, taxes, interest rates, commodity abundance, and the like get lumped into an index of 'leading economic indicators'; the prices of a diverse basket of goods and services are aggregated into a 'consumer price index'; and the attitudes of millions of individuals making personal economic decisions about spending, saving, investing, or retiring is represented with a 'consumer confidence index'. The behavior of the economy as a whole is then presumed to be the outcome of some quantitative relationships divined by economists that link the balance of trade, the consumer price index, the consumer confidence index, and several other aggregate indices. The quantities to be aggregated, and the relationships between them, constitute an economic model. Different economic models arise because different economists make differing assumptions about how the components of the economy relate to and interact with each other. In the USA, there are many models to choose from and to evaluate: the Congressional Budget Office Model, the Wharton School Model, the Senate Fiscal Agency Model, the University of Michigan Model. Each model incorporates the judgments and perspectives of its creators, and some will inevitably prove to be more insightful and prescient than others.

CONCEPTUAL MODELS

Perhaps the simplest models are conceptual models. A conceptual model is a mental image of a system, its components, its interactions. It lays the foundation for more elaborate models, such as physical or numerical models. A conceptual model provides a framework in which to think about the workings of a system or about problem solving in general. An ensuing operational model can be no better than its underlying conceptualization.

A familiar conceptual model often developed in introductory economics courses is the Law of Supply and Demand. The concept of supply and demand relates production and consumption through the mechanism of pricing. If there is an abundance of production of a particular product or commodity, in excess of the demand for that product at the existing price, then the vendor of the product may choose to lower the price to make the item more attractive, or the producer may choose to curtail production because the item is not selling well. Conversely, if an item sells so well at the current going price that the merchant cannot keep the item on the shelf, she may be tempted to raise the price to take advantage of the strong demand. Alternatively, the manufacturer of the item may elect to increase production because of the popularity of the product. Or a competitor, seeing an opportunity to open a new product line, may decide to begin production. In an ideal free market, adjustments in production, consumption, and price take place continuously to maintain an equilibrium between these factors. As a simple conceptual model, supply and demand has in broad and general terms described the operation of a free-market economy reasonably well.

If, however, a conceptual model is too limited in its vision, or if it incorporates false assumptions, then its ability to predict the behavior of a system will be limited and/or misleading. Supply and demand may work well in an ideally free market, with many suppliers and many consumers. But it can be a rather poor description of market interactions when there are monopolies in the supply side, inadequate channels of distribution, or prices that are controlled by a regulatory agency or supported by subsidies. The issue of agricultural subsidies in the European Economic Community has been perpetually vexing as the countries of Europe struggle to balance free trade with the need to maintain a strong agricultural sector in their economies.

For decades, the electric utility industry in the USA has been characterized by all of these departures from an ideal supply and demand economy. Deregulation of the utilities, already underway or planned in several states, will alter the playing field in ways untested, but the initial experience in the state of California has provided dramatic evidence that deregulation does not always go according to script.[1] Deregulation of the telecommunications industry has been underway for more than a decade, with extraordinary changes in the way we transmit and receive sounds, images, information, and data, many of which were unanticipated at the time of deregulation. To be

---

[1] I discuss utility deregulation in California briefly in Chapter 9, in the context of predictions that go awry.

sure, there are many surprises yet to unfold in that aspect of information technology.

At a very basic level, the deregulation of the airline industry in the 1980s assumed that increased competition would serve customers better through lower prices and better, more convenient service. But deregulation created a new framework for airlines to minimize their operational costs. It very quickly led to the establishment of the hub and spoke system of airline routes, with each airline selecting a few hub cities, in which they became the dominant carrier. In some situations, the market dominance has been tantamount to a virtual monopoly, with an attendant decline in service and little incentive to decrease fares. When smaller airlines attempt to capture part of the market in the hub cities, the larger established airlines temporarily cut prices to drive the smaller competitors out of the market. Customers soon have no other carrier to turn to, so there is little need for the established airline to worry about defections. The conceptual model of supply and demand did not prove to be a very good description of the way the airline industry adapted to and took advantage of the deregulation.

The minimization of operational costs also took its toll on 11 September 2001. The airlines, treating airport security as a 'cost center' to be managed, had contracted with private-sector security corporations who, in turn, hired screeners at the minimum wage, without fringe benefits, and set them to work with minimal training. The World Trade Center tragedy was in part attributable to lax airport security. The subsequent decision by the U. S. Congress and the President to federalize airport security was an acknowledgment of the failure of the airline companies to provide adequate security.

INCOMPLETE CONCEPTUALIZATIONS
In science, models can also be good or bad, depending on how well the scientist has conceptualized the system: how much insight he or she has in understanding the workings of a complex system. Let us look

at some examples of incomplete or flawed conceptualizations drawn from the history of science, and how revisions of these inadequate conceptualizations led to progress.

Authors of geology textbooks always like to cite one well-known nineteenth century approach to the problem of determining the age of the Earth as an example of an incomplete and, therefore, misleading conceptualization. Determining the age of the Earth has long been a central topic in the geological sciences; today the most reliable estimate of Earth's antiquity comes from methods that use the decay of radioactive elements, for example the decay of uranium into lead, by various nuclear processes that proceed at a regular and well-known rate. However, radioactivity was not discovered until the end of the nineteenth century; prior to that discovery, scientists used other approaches to estimate the age of the Earth. The incompletely conceptualized and ultimately erroneous estimate was made by the Scottish physicist William Thompson, better known by his peerage title of Lord Kelvin.

Kelvin's reasoning went along these lines: the Earth was endowed with a certain amount of heat at the time of its origin and has been cooling off ever since. If one could determine how much heat it originally had, and had a good understanding of how it lost heat, this could lead to an estimate of how long the cooling had been going on, or equivalently how much time had passed since the Earth was formed. An analogy to Kelvin's methodology would be a bathtub full of water; at a given time the plug is pulled and the water begins to drain. At some time later, the water level has fallen, and if one knows the rate that water leaves the tub through the drain, one can calculate from the reduced water level when the plug had been pulled.

Kelvin reasoned that an upper limit to the amount of heat the Earth was originally endowed with could be determined from the temperature at which rocks melted, a temperature that could be measured at the site of volcanic eruptions. This logic implicitly assumed that the Earth had originally formed in a molten condition but soon thereafter solidified into the rock we see today . He next argued that the

Earth would lose heat by a process called thermal conduction, a style of heat transfer that takes place in solids such as the rocks that make up the Earth's crust today. His conceptualization of the process was not fundamentally different from the way a rock around a campfire slowly cools after the campfire goes out. In a domestic context, conduction is the heat transfer process by which a teacup warms when hot tea is poured into it.

Kelvin knew that the rate of cooling would not be uniform through time; Earth would cool more rapidly shortly after formation, and more slowly as time went on. So, if he could determine the rate at which Earth was losing heat at the present time, he would be able to tell how long the cooling had been taking place. He proceeded to make careful temperature measurements in underground mines to determine the present-day rate of heat loss, which he then used to calculate how long the Earth had been cooling. His estimate of several tens of millions of years, while much longer than the biblical estimate of a few thousand years, fell far short of the eons of time that geologists estimated were necessary to account for the geologic landscape they observed and biologists thought necessary for evolution of species. Thus was set in place one of the fiercest intellectual battles of the nineteenth century, with geologists and Charles Darwin in one corner arguing for great antiquity and Kelvin, the physicist, arguing for a considerably younger Earth.

The debate continued without resolution, each camp believing the other wrong, until the discovery of radioactivity at the very end of the nineteenth century led to the unraveling of the Kelvin argument. Radioactive decay of unstable elements, first observed and described by the French physicist Becquerel in 1896, is a source of heat, one that we are familiar with in the operation of nuclear power plants, where the energy of decay is transformed into heat that is used to produce the steam that drives the turbines that generate electricity. Of what significance was radioactive decay to the debate about the age of the Earth? Kelvin's entire argument hinged on there being only one source of heat to be lost, the endowment of heat the Earth had inherited from its original molten condition. Because rocks contain small amounts of radioactive elements, their decay was providing new heat at the same time that the original heat was being lost. That would, of course, mean that it would take longer to cool than if there were no extra sources of heat. Returning to the analogy of determining how long a bathtub had been draining, it would be as if the experiment had been performed without realizing that the tub's faucets were open, adding water to the system at the same time as it was being lost through the drain.

Kelvin had not miscalculated, but he had incompletely conceptualized the problem. He had no way of knowing that the Earth's heat budget had not just its inheritance but also a source of income, the heat of radioactive decay. Because of the existence of the yet-to-be-discovered radioactivity, and the presence of radioactive elements in the rocks of the Earth, Kelvin's determination of the age of the Earth was grossly inaccurate. It is ironic that the very phenomenon, radioactivity, that was the undoing of his approach to the age of the Earth, would later prove to be central to the currently most reliable method to determine when Earth was born. As mentioned earlier, this approach is based on the steady decay of a radioactive element (the 'parent') into a stable element (the 'daughter'). Older rocks have less parent and more daughter than younger rocks. Uranium parents and their daughters of lead have shown that Earth is slightly older than 4.5 billion years.

FLAWED CONCEPTUALIZATIONS

Flawed conceptualizations abound in the history of science. Some are particularly instructive in illustrating how misconception impedes understanding, and consequently contributes to uncertainty. First let us look at the famous reframing of the structure and dynamics of our planetary system by Nicholas Copernicus (1473–1543), completed around 1530 but not published until just shortly before his death.

The human fascination with the physical universe that surrounds us is ancient. The Babylonians of more than three millennia ago were keen observers of the nighttime sky. They noted that most of the points of light remained fixed relative to one another, in the geometric patterns we now call constellations. But a few bright spots were moving across the patterns, and the astronomers of Babylon called them 'wild sheep'. Today we recognize the wild sheep as our near neighbors in space and call them planets after the Greek word signifying 'wanderer'. The Greeks added their own observations to those of the Persians and Babylonians and developed a picture of the celestial bodies that placed the Earth at the very center of the universe. In this 'geocentric' picture, the Sun, Moon and planets moved round the Earth, like handmaidens in service to a monarch, and the constellations remained in the background, as the inhabitants of the far reaches of the realm. Later, this view was set forth in the writings of Claudius Ptolemy, a second century astronomer and geographer in Alexandria, a center of intellectual life at the time and today a suburb of Cairo. The Ptolemaic view of the cosmic geography had immense philosophical and religious attractiveness. The geocentric cosmos was seen to be the exquisite work of God, who created humans on Earth as the centerpiece of the universe and created everything else as a backdrop.

The Ptolemaic view influenced the thinking of virtually all the European and Mediterranean world until that conceptualization was challenged by Copernicus. Schooled in the conventional geocentric perspective of the cosmos, Copernicus found it less than satisfying. His intuition led him to believe that it was the Sun, the great source of heat and light, that occupied the center of the planetary system, and that the Earth was just another planet that orbited about the Sun. From such a perspective, the apparent daily motion of the Sun and Moon and the annual shift of the constellations across the sky could easily be explained in terms of the Earth rotating on an axis daily, and orbiting about the Sun yearly. As a side benefit, it also allowed for a simpler description of the motion of the planets as seen from Earth than was possible with the geocentric system. Copernicus believed that a simple explanation was more appealing than a complex one, a principle of parsimony enunciated two centuries earlier by William of

Occam, and known popularly today as 'Occam's razor'. (Occam, too, crossed the establishment of his time by expressing iconoclastic views and found himself in disfavor for departing from the conventional wisdom.)

But what a challenge Copernicus had made! In effect it amounted to a serious demotion for Earth, from being the lofty monarch of the realm down to being one of a rag-tag team of hangers-on in the royal court. It was a reassignment from the center of the universe to third rock from the Sun. It was an affront to the religious and philosophical establishment that had grown accustomed to the importance, power, and wealth that accompanied being the official interpreters of God's great scheme of creation.

Ideas do have a life of their own, however, whether or not they are widely believed when first set forth. The concept of a Sun-centered planetary system found other fertile minds in which to grow. The ideas of Copernicus motivated Tycho Brahe (1546–1601) to make further painstaking observations of the motions of the planets, and Johannes Kepler (1571–1630) discovered certain regularities in Tycho's data that he summarized in his famous three laws of planetary motion. Isaac Newton (1643–1727) developed an elegant mathematical description of the forces and interactions that must exist to yield Kepler's planetary orbits about the Sun. Today, we call that interaction gravity, and the planetary system is known as the *solar* system, in recognition of the central position of the Sun.

The significance of the Copernican revolution, the replacement of the geocentric or Earth-centered conceptualization with the heliocentric or Sun-centered framework, is that understanding of a complex phenomenon is dependent upon a proper conceptualization of the system being studied. No deep understanding of the structure and dynamics of the solar system was possible until the notion that the Earth was at the center of the solar system was abandoned. Once that had been achieved, however, once the shackles of misconceptualization had been broken, enormous strides were taken one after another. Today we benefit from the daily weather images beamed from

Earth-orbiting satellites, we share in the excitement of seeing astronauts on the Moon or a robotic rover sending us pictures from Mars, and we are astounded with the precision of ballistic missile trajectories. All of these became possible because of advances in understanding enabled by a proper conceptualization of the planetary system.

CONTINENTS ADRIFT

Another example that illustrates how a flawed conceptualization impedes progress can be drawn from the history of the geological sciences. The concept of continents drifting over the surface of Earth throughout geologic time drew its earliest adherents from geographers who noted the great similarity between the configuration of the west coast of Africa and the east coast of South America. To the imaginative eye, the two continents could fit neatly together like pieces of a jigsaw puzzle. The geological inference drawn from that apparent similarity was that South America and Africa were once joined, only to split apart and separate from each other sometime in the distant geologic past. Sir Francis Bacon commented upon this remarkable geography in the early seventeenth century, as did others in the eighteenth century. By the end of the nineteenth century, geologists had discovered that at corresponding points along the two continental margins the rocks and fossils were very similar, in effect showing that not only did the puzzle pieces fit together but also the picture was continuous across the fit. Early in the twentieth century, the geographic and geologic evidence was summarized and interpreted by the German scientist Alfred Wegener in a now-famous book titled *The Origins of Continents and Oceans*.[2] Wegener presented a compelling case that the continents had once been joined and had since drifted apart.

Wegener's concept, while particularly attractive to those who had studied southern hemisphere geology, did not sit well with

---

[2] Wegener's book was first published in German in 1915 (*Die Enstehung der Kontinente und Ozeane.*). An accessible English translation was published by Dover Publications, New York, in 1966.

another group of Earth scientists, the geophysicists who studied physical properties of rocks. They pointed out that between South America and Africa, hidden by the waters of the Atlantic Ocean, lay some three thousand miles of solid rock constituting the crust of the Earth beneath the ocean. The geophysicists conceptualized continental drift as a process analogous to ships plowing through the seas. How, they asked, could the continents plow their way through this barrier of rock? They argued that the rocks forming the floor of the ocean were much too stiff to permit such motion. Because of the lack of an obvious mechanism by which continents could drift, the geographic and geologic evidence that suggested the continents *had* moved around over the Earth was ignored for decades. The jig-saw puzzle fit was trivially dismissed as nothing more significant than another well-known geographic oddity, that Italy looks like a boot. And as for the similar rocks and fossils on either side of the Atlantic, well, they could have resulted from similar processes and parallel evolution.

This state of affairs persisted until the mid-1960s, when a new concept called plate tectonics emerged, one that permitted a re-evaluation of the old arguments about continental drift. The new concept accepted that continents could not plow their way through the oceanic crustal rocks. Instead, it proposed that both the continents *and the oceanic rock surrounding them* were moving together, much like a log frozen into a sheet of moving ice. As sheets of ice with their log 'continents' separate from each other along major fissures, the exposed water between them freezes, creating new 'rock' in the gap between.

This revisionist view of the Earth was stimulated by important new observations about the age of the rock making up the ocean floor. The data were actually not so new, having been gathered for military purposes associated with submarine warfare during World War II and the cold war thereafter. When these data were released to the scientific community, they revealed that the rocks of the ocean floor were all younger than most of the rocks of the continents, just as the new ice forming between separating ice sheets on a lake is younger than the ice sheets themselves. Thus the abundant evidence for continental drift that geologists had assembled no longer had to be set aside because of the argument that continental drift was a mechanical impossibility. The mobility of continents had been rejected because of the misconception that continents were actively propelling themselves through the rigid crust of the Earth. Once it was realized that continents were passive passengers along for the ride, the geologic evidence was quickly recognized as valid.

Just as the revolutionary ideas of Copernicus led to extraordinary advances in understanding the solar system, so also did the Earth sciences benefit extraordinarily from the newly developed plate tectonic concept of Earth dynamics. The new view of the Earth, made possible by new oceanographic data, stimulated a torrent of creative thinking about such practical problems as the generation of petroleum and natural gas and the formation of ore deposits, about the nature of seismic and volcanic hazards, and about factors affecting long-term climate change. A former view had been abandoned because new observations forced a rethinking, a new conceptualization that could accommodate the recent observations and at the same time reconcile them with the older evidence. Plate tectonics was off and running, and Earth science was forever transformed.

IN AND OUT OF RUTS

Seldom is the idea of being in a rut thought of as a positive situation. Usually we associate that condition with frustration, discontent, and an impoverishment of new ways of thinking about problems. We are implored to 'think outside of the box', to break the barriers that limit the ways we conceptualize problems. Reginald V. Jones, a professor of physics at the University of Aberdeen in Scotland, taught me much about the importance of avoiding conceptual ruts, and of the value of recognizing ruts as soon as possible so that one can get out before the rut is too deep and too much time has been wasted.

As a young man in Britain during World War II, Jones played an important role in scientific research and intelligence as applied to the war effort. In his book *Most Secret War*[3] he recounts a story about escaping from ruts. One of his team's early assignments was to develop an aircraft that could not be detected by radar. Already, radar was playing an important role in providing early warning of impending enemy bombing raids on the British Isles, as well as warning the Germans when Allied planes were making their way to European targets. If Jones could devise a way to make aircraft invisible to radar, then the Allies would have a distinct advantage in reaching their targets unmolested by Luftwaffe interceptors. Today this very same concept has been implemented in the family of 'Stealth' aircraft operated by the US Air Force.

Jones and his team tried all kinds of technological trick, all to no avail. An aircraft made of wood, special geometries for the wings and fuselage, rubber coatings – but nothing could prevent the strong radar reflection from the big metal engine. They tried and tried without success. Jones realized they were in a rut, that they were not thinking outside the box. Finally a solution came to him: "How", he asked, "do you hide a grain of sand?" The answer was straightforward: you put it on the beach, amidst an infinity of other grains of sand. Translating this to the business of disguising airplanes, the solution was not to suppress a single radar signal, but rather to create a million radar signals. Thus was born the strategy of dumping shredded metallic foil, such as we use in wrapping food for storage in the refrigerator, from a decoy aircraft. To the enemy radar, each piece of metallic foil looked like an incoming raider, and the ensemble of shreds appeared as a virtual armada of bombers and fighters on its way to wreak destruction. The defensive interceptors took off to engage this overwhelming force, only to reach the drop area and find no planes whatsoever. Meanwhile, a much smaller group of real bombers went in a different direction over enemy territory and delivered its high explosive cargo virtually unmolested. The deception lasted only a few weeks, but considerable damage was inflicted in that period.

[3] Reginald. V. Jones, *Most Secret War: British Scientific Intelligence 1939–1945.* Hamish Hamilton, London, 1978. Also reprinted in paperback by Wordsworth Editions, 1998.

In order to be effective in thinking 'outside the box', however, we must first be aware that we are indeed 'inside the box', constrained in the ways we think about problems. What are some of these often-subtle constraints? Let's start with wishful thinking. When affairs, social or financial, seem not to be going well, we often say to ourselves "things will get better". When an investor's favorite stock has undergone a long and painful slide, at many stages of the descent he or she will have argued that things cannot get worse, that the stock price has bottomed out and will soon be on the road to recovery. In such cases, the wishbone governed while the backbone buckled. This attitude is frequently the response to many emergent environmental problems. We recognize that air quality has deteriorated, that the roads and highways are congested, the water at the beach is unsafe for swimming... but we say to ourselves things will get better. However, hardly any problem is fortuitously solved by spontaneous remediation, or by simply wishing it away. Changes usually occur only after the problem is engaged, analyzed, and acted upon.

Albert Einstein observed "You cannot solve current problems with current thinking. Current problems are the result of current thinking." Current thinking is just another name for conventional wisdom. As a barrier to problem solving, conventional wisdom appears in a number of different forms: we have always done it this way; we tried that before; this remedy will cost too much; that path surely leads to failure. Conventional wisdom is repetition without examination; it is an acceptance of current perspectives without asking whether the foundations that support such perspectives are themselves undergoing change. Rejecting or challenging the conventional wisdom is an act that replaces apparent certainty with disconcerting uncertainty. Had Copernicus or Wegener simply accepted the conventional wisdom of their time, where might we be today in understanding how the solar system and Earth work?

Ideology is a special form of conventional wisdom that is commonly backed by some important clout. Governments and religious institutions sometimes establish conventional wisdom and then protect it from routine challenges. Behind this ideological defense one can often find wishful thinking. The Ptolemaic concept of an Earth-centered planetary system had acquired a pre-Copernican ideological status because the religious and secular institutions of the day wanted it to be true. Centuries later, the science of genetics in the Soviet Union was set back because of the imposition of an ideological box on scientific thinking. A key player in the establishment of the misguided genetics was Trofim Lysenko, a biologist and agronomist conducting research at an agricultural experimental station in the Ukraine. He claimed that wheat seeds could be made to yield rye under proper environmental conditions, and that the ability could be transferred to subsequent generations of grain. This was particularly attractive to the Soviet ideology that humans were socially malleable, and that good socialist attitudes could not only be taught but also inherited. The Soviets in power *wanted* Lysenko's agricultural ideas to be true, and for a time barred the teaching of and research in mainstream genetics as it was developing elsewhere in the world. There was an allegiance to ideology rather than to science.

In these early years of the twenty-first century, much of the opposition to recognizing that global climate change is taking place, and to acknowledging that humans are playing a significant role in causing the change, is based both on wishful thinking and on conventional wisdom. The fossil fuel and transportation industries simply do not want to believe that something extraordinary is taking place that may force them to reconfigure, or even worse to abandon, what has been a century-long success story. That we have utilized fossil fuels for more than a century to them signifies "we have always used fossil fuels, and always will". This perspective, as long as it is prevalent and persuasive, will lead us into an ever-deepening rut.

Wishful thinking is seldom a successful long-term strategy, and sooner or later reality will force an acknowledgment of change. The stone age of human history came to an end not because humans ran out of stones, but because humans learned to make better tools from metal. Similarly the fossil fuel stage of human development will also end well before we have exhausted the fossil fuels. In the long sweep of human history, both past and future, the era of fossil fuel reliance will be seen as a temporary crutch during the interval when humans learned how to harness and concentrate the virtually endless radiative energy delivered directly to Earth by the Sun. Fossil fuels such as coal and petroleum are, in fact, nature's somewhat clumsy and inefficient products in which is stored the solar energy of the past, in the form of fossilized vegetation and microorganisms, appropriately decomposed and recomposed through cooking while buried in the Earth's crust.

PHYSICAL MODELS

As a child I played with Lincoln Logs™, Tinker-Toys™, and Erector Sets™, constructing an infinite variety of log cabins, windmills, and metal-girdered bridges. A little later, I turned to model airplanes, painstakingly cutting out balsa wood spars to form wings and a fuselage and then covering them with a tissue paper skin. With a propeller and rubber band for power, the miniature aircraft took to the air. All of these childhood items were simplified versions of real-world constructs, scaled down physical models of the actual edifices of the real-world. As physical models they represented many of the characteristics of the real-world structures, but in other significant ways they departed from their large-scale counterparts. A simple model airplane used paper instead of fabric or metal for its skin, elastic energy from a rubber band instead of chemical energy from liquid fuels for its power, and had no pilot or controls. If one tried to build a real airplane that mimicked the model, one quickly would recognize the inadequacies of the model. In fact, the model airplane was a model only of the simple aerodynamic principles of flight, not of an airplane that required control through takeoff, flight and landing, that needed cargo and passenger space, and fuel for long-distance journeys.

Physical models have advanced well beyond my childhood experiences. Automobile manufacturers and aeronautical research centers still utilize wind tunnels to study turbulence generated by various vehicle and aircraft designs, and naval architects have large tanks and basins in which they tow or propel different hull designs to learn how to improve speed, fuel efficiency, and stability under various wave and wind conditions. Physical models continue to play an important role in experimentation, and I will have more to say about them in the next chapter. But the success or failure of physical model studies depends on how well an experiment done at one scale, and under conditions not exactly the same as in the real world, will extrapolate into an actual plane or ship, in the air or on the open ocean under real weather conditions.

NUMERICAL MODELS

Numerical models comprise a set of mathematical equations that describe the functioning of a system and that can generate quantitative predictions about its behavior. When the calculations are so complex that they require a computer to carry them out, then the numerical model is sometimes called a computer model. But whether the calculations are done by hand, with an old fashioned adding machine or a hand calculator, or with a super-computer, at the heart of all numerical models are the equations that quantitatively link the components and processes of the system. Lord Kelvin calculated the age of the Earth by using a solution to the differential equation of heat conduction; that solution enabled him to quantify the rate at which a sphere loses heat from an initially hot condition and thereby calculate how long Earth has been cooling.

To get a flavor for numerical models, let us start with something simpler and more intuitive. Take, for example, a model of your savings account, in which you make a deposit each month and the funds in the account earn interest at a given rate that is compounded monthly. There are simple mathematical rules that enable you to calculate the value of your savings at any future time. Those rules are, in effect, a numerical model of your savings account, a model that represents the deposits and the compounding over time. It could be made more complex if you wanted to make periodic withdrawals to pay for your auto insurance or rent.

The schedule of mortgage payments that will amortize your home loan over a fixed number of years likewise is the output of a simple numerical model. The model takes the amount you borrowed, the interest rate you have negotiated with the lender, and the number of years over which you wish to spread the payments, and it then employs straightforward mathematics to tell you what size monthly payment you must make. Additionally, it tells you how much of each monthly payment goes to pay interest, and how much goes to reduce the loan principal. At any time in the life of the loan, the model will tell you how much principal is left to be repaid, should you decide to pay it off with an inheritance received from Aunt Jenny, or a large bonus from the office.

SOCIAL SECURITY

Another model that Americans are familiar with is that which underlies the US Social Security system. This is an economic model that has many aspects in common with an individual savings account, but it also has some characteristics that are different and more complex. One difference is that the system anticipates not only savings during an individual's working years but also payouts in the later retirement years, with the payout tied to the level of accumulated savings. A second difference is that there is no time limit to the payout; the benefits will continue as long as one lives. By contrast, an individual savings account will permit withdrawals only as long as a balance remains. A third and important difference is that the Social Security system has millions of participants who pool their savings, and at the same time the system is making payouts from the savings pool to seniors. The reason I call this a model is that the operators of the Social Security system, the Social Security administration, must estimate for any given future year how many people will be working and contributing and how many people will be retired and receiving payments. And they must estimate how long people are going to live. Moreover, they must make such estimates many decades into the future. It is not difficult to understand why there is uncertainty associated with the financial viability of the Social Security system.

When the Social Security system was introduced in the 1930s, in order to determine how much each worker would have to contribute to the savings pool, the administrators made informed guesses about family size, immigration trends, employment levels, retirement ages, life expectancy, and interest rates long into the future. Were those estimates on the mark? To anyone who contemplates such issues professionally – demographers, economists, actuaries – it came as no surprise that the estimates used initially in the calculation did not exactly produce the reality that unfolded in the twentieth century. What had been predicted in the 1930s was decidedly different than what actually evolved over the subsequent seven decades.

At the end of the twentieth century, it became apparent that within another few decades, contributions to the Social Security system would begin to lag withdrawals, in part because families had become smaller and provided fewer workers, and in part because people were living longer. In 1935, only 6.5 million Americans, some 5% of the population, were over the age of 65. In 2000, the numbers had increased to 35 million, making up 13% of the nation. Half of all the people who had ever reached age 65 in the USA were alive at the beginning of the twenty-first century. Just as with weather predictions, the farther into the future one tries to project, the more tentative and uncertain the prediction becomes.

Weather forecasts are reasonably accurate for four or five days into the future, on a rolling basis. This success comes because every day the calculation for the future is adjusted for whatever small differences have appeared between prediction and reality. In effect, the calculation has undergone a slight course correction, and the calculation is back on track for a good prediction for the next day. The Social Security calculation likewise has undergone adjustments. As shortfalls began to loom as a future possibility, the Social Security administrators, with Congressional approval, every few years increased the base amount of salary to be taxed for contributions to the pool. Recently, in recognition of the increase in life expectancy that has evolved over the twentieth century, the age at which full benefits can be withdrawn from the savings pool is slowly being increased from 65 to 67. Even today, there are discussions about whether the investment earnings of the savings pool might be enhanced to bring the savings pool and the benefits into closer alignment.

An important lesson of this discussion of the quantitative model of the Social Security system is that one could not, at the time of initiating the system, anticipate all of the eventualities that made the predictive model increasingly inadequate as a description of the real world. However, uncertainty about the future did not stop the introduction of the system, and mid-course corrections over the years have modified the system to extend its viability. Decisions must always be made in the face of uncertainty, and for all of the inadequacies that may be ascribed to the Social Security system by its detractors, there has been and continues to be strong public support for the basic concept of social security and appreciation for its establishment seven decades ago, in a climate of great uncertainty.

Models invite modification so that they better represent the real world they imitate. And the model improvement comes from experiments wherein the model is confronted with varying conditions and stimuli, and its response or reaction is carefully observed. A model aircraft that crashes under certain conditions of turbulence will surely be returned to the drawing board for modification. In the model of the Social Security system, one can run 'what if' experiments to see what additional revenues would be needed to address the effects of extending life expectancy by five years over the next two decades. The boundary between models and experiments is almost transparent, and so let us arbitrarily call an end to this chapter on models and move on to discuss experiments.

8 Let's see what happens if...

No man really becomes a fool until he stops asking questions.

Charles Steinmetz

The very term 'experiment' implies uncertainty, because why would one want to conduct an experiment if the outcome is certain? The goal of experimentation is to learn something new about a system, something that is unknown, or only poorly understood. Experiments are a natural outgrowth of models, because a model, whether conceptual, physical or numerical, will always be a simplified representation of a system, and experiments with this model help us to understand its strengths and weaknesses. In terms of consequences, the simplifications embodied in the model may not matter under many circumstances, but then along comes the special situation when the model becomes vulnerable. In this context, models invite experiments that put them to the test, in a process of validation.

Just as model building begins with a concept, a mental image of how something is constructed or functions, so also does experimentation begin with what are called 'thought experiments'. These experiments are mental forays that explore the consequences of assumptions or possible paths of action. Albert Einstein was a firm advocate of thought experiments; many of his early concepts about relativity stemmed from his attempts to visualize how the universe would appear if he were to hitch a ride on a beam of light.

Experiments essentially pose questions and seek answers. A good experiment provides an unambiguous answer to a well-posed question. A curious boy might formulate an experiment to answer the question "What is the acceleration of a falling body?" In his experimental design, he envisions dropping a bowling ball from the top floor of a tall building. Prior to the drop, he will station his friends at windows on each lower floor with stopwatches to note the time when the ball passed each window. An analysis of how the time interval between floors became shorter as the ball fell would reveal how fast the ball picked up speed under the influence of gravity. Alternatively, the experiment could be reframed as a test of a hypothesis or prediction: Sir Isaac Newton's second law of motion predicts that the acceleration of a falling body should be equal to twice the distance traveled divided by the square of the travel time. If the observations deviated significantly from this prediction, then the experimenter might question the validity of Newton's second law, or try the experiment over again to see if it gave the same result. The latter strategy is a time-honored way by which scientists test each other's work.

Other experiments might not be quite so simple or straightforward. A fisheries biologist might ask "What effect will agricultural pesticides in lake water have on the reproductive vigor of small-mouth bass?" She would then go about setting up an experiment in tanks at the fish hatchery, with each tank having a different concentration of pesticide. Monitoring the number and viability of the fish progeny from each tank might provide some relevant data. But is an experiment in a tank the same as the 'natural experiment' taking place in a nearby lake surrounded by intensive agriculture? Might factors other than pesticide runoff, such as acid rain and airborne mercury from an upwind power generating plant, also affect fish reproduction in the lake? Might the presence of and competition with other fish species in the lake influence the reproductive success of the bass? Might reproductive success depend on the annual range of temperature the lake experienced? What influence does the age distribution of the fish in the lake have on their fecundity?

The biologist may answer the narrow question posed with her tank experiments at the hatchery but leave unanswered the larger question of what is really happening in the natural setting. The natural setting is undergoing a natural experiment in which conditions, influenced but not controlled by human activity, are changing and the system is responding. Clearly, it is difficult to simulate nature in the laboratory. The uncertainties associated with the results of the fish hatchery experiment are larger than just the uncertainty of the measurements in the experimental tank. The greater uncertainty derives from whether the experiment is a good representation of what is happening in nature.

The way an experimenter goes about designing an experiment is crucial to the ultimate success or failure of an experiment. Frequently, an experiment is designed to test a preconceived idea about how a system behaves under various tweakings. Earlier we called this a conceptual model. In the fish hatchery experiment, the hypothesis may have been that pesticides in the water in relatively low concentrations had no effect on fish reproduction, and the single experimental variable was the concentration of the pesticide.

Preconceived ideas about the behavior of a system, while almost a necessity in designing an experiment, can also be an impediment to executing it and interpreting the outcome. The medical and pharmaceutical sciences frequently use an experimental procedure known as a double-blind protocol, which is designed to minimize the entry of bias in the performance of an experiment. In the testing of a new drug, a group of patients will be selected, half of which will receive the drug and the other half a placebo.[1] However, in a double-blind experiment neither the patients nor the researchers administering the test will know which patient has received the drug and which the placebo. Of course, through careful prior coding of both patients and treatments the knowledge of who received what will be retained, so that after a monitoring period the efficacy of the drug can be separated from the so-called placebo effect.[2] Such a procedure substantially reduces the biasing of the outcome of an experiment with patients who were sadly aware that they had been given a placebo, or by researchers who offered encouragement to patients who had received the experimental drug.

[1] A placebo is a substance that has no active ingredient that could affect a patient's condition.
[2] The reality of the placebo effect has recently been called into question. See the *New York Times*, 24 May 2001.

While preconception is almost unavoidable in experimental design, it can also be a stumbling block in evaluating the results. What if the experiment revealed a result quite different from the preconception? The standard scientific lore is that the experimenter will then be persuaded by the evidence that conflicts with the prior hypothesis and will alter his conception of the system. In the real world, however, there is sometimes a reluctance to admit defeat, to abandon one's hypothesis so easily. There is a tenacity with which some experimenters hold on to their initial ideas, even in the face of experimental evidence that suggests, or even shouts, for a new conception of the way the system works.[3] In science, there is an important admonition: "Don't fall in love with your own hypothesis".

INSIDE THE EARTH

Some experiments do not provide a straightforward answer, they simply narrow the range of possible answers. In such experiments, the significant outcome is the elimination of some proposed answers to the question. For example, children often ask, and geologists struggle to answer, "What is the inside of the Earth made of?" With apologies to Jules Verne, a journey to the center of the Earth is not a real possibility. The deepest mines and the deepest boreholes reach depths into the Earth that are in reality only superficial pinpricks, and the samples of rock we obtain from them represent the material in only the outer skin of the Earth. Even the samples of the interior that nature provides us through volcanic eruptions come from only the outermost hundred miles or so, whereas the center of the planet lies almost 4,000 miles below the surface. The vast interior remains inaccessible to direct sampling by both humans and nature.

All is not hopeless however. We learn something about the interior from waves (vibrations) generated by earthquakes. These waves travel through Earth's interior, and when they once again resurface

---

[3] Ways in which preconception influences the interpretation of experiments is discussed extensively by Harry Collins and Trevor Pinch in *The Golem: What You Should Know About Science*, 2nd edition, Cambridge University Press, 1998, 192 pp.

they tell us something about the physical properties of the materials they encountered along the way. These properties include the compressibility, the rigidity and the density. The task that the geologist then faces is to determine what materials have those properties, under the pressure and temperature conditions encountered along the path through the interior. Might the bulk of the Earth be made of granite or limestone, rocks that we find at the surface?

Experimentation with various candidate rocks, squeezing them in giant presses and heating them several thousand degrees to simulate conditions in the interior, helps to determine whether their properties match what earthquake waves encountered on their deep journey through the Earth. After such experiments, granite and limestone are easily dismissed as candidates. No matter how one tortures these rocks with high pressures and temperatures, they do not display the properties of the deep interior. In particular, they never get squeezed so tightly that they attain the density of materials even half way through the Earth. But another type of rock called peridotite, which is not very common at the surface, fares rather well in matching the properties of the outer half of the Earth.

However, no rock can satisfy the characteristics observed for the deepest interior. For that region we must abandon rocks as candidate materials and look for something entirely different. Heavy metals such as iron fit the bill at those depths where rocks no longer meet the test criteria. And so the broad-brush picture of Earth that these experiments suggest is a rocky outer half (not principally granite or limestone, however) enveloping a metal core occupying the inner half.

So do we really know what the inside of the Earth is made of? No, we do not. But the experiments are definitive in defining what Earth is *not* made of: except for a very thin veneer at the surface, the rocky part constituting the outer half of the planet cannot be largely granite or limestone and the dense inner part cannot be rock at all. The materials that pass the experimental tests qualify as candidates for what these regions might be made of, but we cannot rule out that there may be other materials which also meet the requirements. Is the composition of Earth therefore uncertain? Yes. From the seismic and laboratory data (and much additional indirect evidence), we can make confident, well-informed statements about the broad composition of Earth, but in many important details we are much less certain.

Laboratory studies of the mechanical behavior of rocks have also been used to try to understand the physical basis of earthquake occurrence. An earthquake occurs when the rocks of the Earth's crust fracture, allowing the formations on either side of the fracture surface (called a fault by geologists) to slip past each other. In the laboratory investigations, small rock samples are squeezed, stretched, or twisted in massive hydraulic devices until they break, and observations are made about how much torture is necessary to break a given type of rock.

Can the laboratory results be readily extrapolated to predict real-world behavior? Not without a lot of caveats. In the real world, the distortion of the Earth's crust by natural processes typically occurs over time scales of thousands and millions of years, whereas in the laboratory, even with a patient experimenter, the squeezing and twisting take place over a few days or months. Materials behave differently when they receive slow torture as in nature, than they do during the relatively fast abuse they receive in the laboratory. In the real world, a fault associated with a major earthquake may rupture over several miles, whereas in a laboratory experiment the typical sample size stressed to failure is only a few inches long. In the much larger natural setting, there is a much greater chance for heterogeneity in rock types to play a significant role in determining where and when a fault might develop, whereas the laboratory sample is likely to be far more homogeneous. It is simply a fact of life for geologists that their attempts to explore geologic processes in the laboratory run head on into the obvious fact that the Earth does not fit easily into a laboratory, and that humans do not have the luxury of time to conduct experiments at the pace of natural processes. Anticipating the behavior of a large natural system from small and simplified laboratory models is a tough business, fraught with uncertainty.

NUMERICAL EXPERIMENTS

The foundations of numerical experiments are numerical models, those quantitative descriptions of the system being investigated. Numerical models that yield daily weather forecasts employ the physical principles of fluid dynamics; numerical models of the national economy employ economic concepts about the interactions of a myriad of financial, political, and demographic variables. Needless to say, if the physical principles or economic concepts are incomplete or incorrect, the corresponding projections of the future will also be tenuous. To the extent that these complex systems are incompletely understood, the future remains uncertain.

Numerical models, because they reside in the bowels of computers, are particularly amenable to running quickly through many experiments that explore the response of the model to a variety of input factors. We call these 'what if' experiments; "*What* will be the savings in interest charges *if* I pay my mortgage off in 20 years instead of 30?" Numerical models, put through their paces in a long series of numerical 'what if' experiments, can define a range of outcomes that are the consequences of many different input scenarios.

In many settings, numerical models have superseded physical models in experimentation because they are cheaper and more versatile, and thereby enable experimentation over a much wider range of variables. One can try many more 'what if' experiments inside a computer than can be undertaken in laboratory experiments with real materials, or with model airplanes in wind tunnels. And some experiments can be done only on computers. We cannot answer through direct physical experimentation a question such as "What would Earth's climate be like if the Atlantic and Pacific Oceans were not separated by the Isthmus of Panama?" But we can create a numerical model of atmospheric and oceanic circulation that will enable experimentation with alternative geographic configurations of oceans and continents.

THE AMERICA'S CUP

Every few years, the yachters of the world take to the seas in a great competition known as the America's Cup race. This race, begun in 1851 as a spirited competition between American and British gentlemen sailors, today has evolved into a worldwide international rivalry involving corporate sponsors and arcane technical rules. Nevertheless, it still boils down to keen sailing skill and a never-ending search for a technical edge in hull and sail design. It is in the latter area, the pursuit of technological advantage, where modeling plays a central role.

Earlier in the history of this race, new hull designs were actually built and tested in the water to see what improvements in speed they yielded. But that proved exorbitantly expensive as a trial-and-error strategy (racing yachts have never been inexpensive toys). Later, reduced-scale models, towed or blown through indoor hydrodynamic testing tanks, replaced full-scale hull development for purposes of experimentation. Today, the arena for model experiments is all high and dry inside fast scientific computers, where the equations of hydrodynamics enable virtual water to interact with a virtual hull. Such investigations are numerical experiments, in contrast to experiments conducted in a laboratory or in a natural setting. Numerical experiments in computers have replaced real hulls in water as the source of design insight. Varying wind and sea conditions are simulated numerically, testing hull and sail configurations to determine which might enable a skilled racer to eke out an advantage in speed. Ultimately, the results of such experiments must face the real world test of racing under natural conditions in the ocean.

TESTING NUCLEAR WEAPONS

An even bolder step in the direction of reliance on computer modeling and experimentation lies in the design of nuclear weapons and an assessment of their shelf life. Since the development of nuclear weapons and their use in bringing World War II to an end, a whole array of new weapons, both bigger and smaller, have been designed and tested to ensure they will perform as expected. Arsenals of many thousands of warheads of varying design have been accumulated by the nuclear powers, a designation that originally included only the USA, the Soviet Union, France and Great Britain, but which today has China, India, Pakistan and perhaps other countries also under the umbrella. So far, except for the two bombs utilized in World War II against Japan, no nuclear weapons have been used in hostilities. Because of this lack of use, some concern has arisen as to whether a weapon would continue to work after sitting unused for long periods of post-assembly time. So, from time to time, the USA, and presumably the other nuclear nations, would detonate a nuclear device in order to determine whether it had degraded in any way since manufacture. In the USA, these tests were conducted underground at the Nevada test site not far from Las Vegas. These tests of off-the-shelf weapons were real-world physical experiments.

The principal nuclear nations have also realized that the risk of nuclear proliferation may pose a greater threat to their security than any incremental security derived from new weapon development. As a result an international treaty to ban the testing of nuclear weapons, both new and old, has been drafted and placed on the international diplomatic table. This treaty, known as the Comprehensive Test Ban Treaty (CTBT), was signed by US President Bill Clinton in 1996, but it has not been ratified by the US Senate. In the brief Senate ratification debates that took place in 2000, a principal issue that arose was how reliable the weapons would be over the years, without benefit of actually testing one periodically to see if all the myriad components continued to work.

Proponents of the treaty have argued that there is no longer a need for the physical development and testing of nuclear weapons, because it is possible to construct computer models of the important physics and engineering aspects of weapons development, models that will obviate the need for real physical experiments with real physical weapons. As for the reliability issue, the workings of a thermonuclear weapon include electrical and mechanical components, and chemical (non-nuclear) explosives, the testing of which are not proscribed by the treaty. So the effects of aging on these components can continue to be investigated, and replacements or improved versions can be installed on stockpiled weapons from time to time.

Although the USA has not ratified the CTBT, it has abided by the provisions of the treaty since 1992, when the USA under the first Bush administration announced a moratorium on nuclear weapons development and testing. Both before and after that moratorium, the reliability of the weapons inventory has been assessed through the Stockpile Stewardship Program of the US Department of Energy. This program emphasizes three activities: (1) the ongoing monitoring of stockpiled weapons to detect defects in the non-nuclear aspects of the weapon, (2) the repair or replacement of components to remedy any defects discovered as a result of the monitoring, and (3) fundamental research into the aging process, to enable the redesign of components as knowledge of their long-term behavior improves.

The investigations on aging include both laboratory experiments and computer modeling of properties affected by slow nuclear decay. Many of the issues associated with the safe storage and containment of spent fuel rods from nuclear power stations also arise in the investigations of aging of nuclear weapons. With a half-life of 24,400 years, the plutonium used in weapons and produced in the reprocessing of power station fuel can affect containment materials over periods far longer than we can observe in laboratory experiments. This gives rise to the need for computer models and simulations.

EXPERIMENTS IN THE SOCIAL SCIENCES
If physical and biological systems such as the global climate system or a tropical forest ecosystem seem daunting in their complexity, the difficulty in conceptualizing and modeling of social systems and human behavior is equally challenging. How does one go about exploring and quantifying the realm of human greed, cooperation, and altruism?

Social scientists have devised simple but very interesting numerical experiments that are executed as computer 'games'. One that explores aspects of greed and fairness is called Split the Pot.[4] This game involves two players. A sum of money, say $100, is put on the table, and one player is asked to decide what fraction of the money he is willing to share with the other player, keeping the rest for himself. If the other player rejects the offer, however, neither player receives anything.

Clearly it is advantageous to make an offer that, even if less than half, is sufficiently large to entice the other player to accept it. Would the other player accept $10, allowing the first player to keep $90? Research has shown that an agreement at this level is very unlikely, even though one might think that any offer should be accepted, because the alternative is to receive nothing. How about a 35/65 split? Maybe. A 50/50 split should clearly be acceptable to the second player, but it has not tested the possibility that he might accept less. What is the best strategy for a player in such a game to maximize his prize? How does the strategy evolve as the game is played many times with the same two players? How does the strategy develop when there are many pairs of players playing simultaneously?

Another computer game that has yielded rich insights into how and under what conditions humans cooperate is called the Prisoners' Dilemma. The game is set in the jailhouse, where two suspects of a robbery are in detention, each being questioned separately. Neither suspect knows what the other is telling the police. Each of the detainees could provide evidence that would implicate the other in the robbery. The police are without substantial evidence and need the help of one or the other of the suspects to break the case open.

Unaware of what the other suspect might be saying, what should each of the suspects do? If neither implicates the other, they will both go free and share the stolen money. If both implicate the other, they both go to jail. But if one implicates the other, while the other remains silent, the silent suspect will go to jail and the other will go free and have all the ill-gotten money for himself. Consequently, there is a

---

[4] Split the Pot is also known as Take It or Leave It, or the Ultimatum Game.

temptation to accuse the other, with a big reward if the other has held fast. But if the other has similar thoughts, and also accuses his partner, both are punished and neither benefits at all. There are only four possible outcomes of this game.

1. Both prisoners keep silent, and share half of the money.
2. Prisoner A implicates Prisoner B, but Prisoner B remains silent. Prisoner A then is released and gets all the money, and Prisoner B goes to jail.
3. Prisoner B implicates Prisoner A, but Prisoner A remains silent. Prisoner B then is released and gets all the money, and Prisoner A goes to jail.
4. Both prisoners implicate each other, and both go to jail.

If we call keeping silent an act of cooperation (C), and implicating the other an act of defection (D), then the four choices can be summarized as C/C, D/C, C/D, and D/D. The dilemma is obvious: silence buys freedom and some money, but only if your partner in crime does the same (C/C). Accusing the other is a gamble with twice the monetary reward, but only if your partner does not implicate you (D/C, C/D). If both implicate each other (D/D), then both are punished without any monetary reward. What would your choice be? Are you cooperative or greedy? And what's your assessment of your partner? Would your behavior differ if you faced the dilemma repeatedly with the same person, versus the situation where the game is played only once?

Multiple encounters, where the players repeatedly test each other in the Prisoners' Dilemma, have led to very interesting observations about strategies to maximize gain in interactions with others. One well-known strategy is called tit-for-tat, a strategy in which the first player always elects not to implicate the other, unless in the previous game the second player has implicated the first. In that situation, the first player gets even by defecting in the very next game, in effect giving tit-for-tat. If the second player backs off from defecting again, the first player also returns to a cooperative stance. This strategy can be characterized as a generally cooperative one, but with instant retaliation if a non-cooperative behavior is encountered. The return to cooperation immediately after a single defection and retaliation event is equivalent to holding no grudges. This game has been played by millions of people in long-running experiments hosted on the Internet,[5] and the results have been extensively analyzed for insights about how cooperation and greed develop and evolve in human interactions. Tit-for-tat, as simple as it is, has proven to be an extremely effective strategy for maximizing long-term gains.

Variants of Prisoners' Dilemma have been developed to examine intra- and extramarital relationships, exploring such issues as whether someone might cheat on their spouse if they thought the affair would not be discovered. Social scientists are now addressing with numerical models and computer games such topics as how people respond to incentives, the benefits of teamwork, why people procrastinate, and what conditions lead to falling crime rates.[6] Some of these models, such as Split the Pot and Prisoners' Dilemma, are remarkably simple, while others involve heavy statistics and mathematics. What they have in common is their focus on the intricacies and subtleties of human behavior.

INADVERTENT EXPERIMENTS

Not all experiments are carefully designed and executed under controlled conditions. Some 'just happen' inadvertently. Yet there is much to be learned from inadvertent 'experiments'. In May of 1990, a container ship carrying Asian cargo to North America encountered very rough seas in the North Pacific, and several steel containers of athletic shoes fell overboard, broke open, and dumped some 80,000 shoes into the sea.[7] The shoes floated nicely and were swept along by the

---

[5] Typing "Prisoners Dilemma" in an Internet search engine will present many opportunities for you to be part of this ongoing experiment.

[6] *The Wall Street Journal*, 27 April 2001; *the New York Times*, 27 November 2001.

[7] Ebbesmeyer, C. E. and Ingraham, W. J., Jr., Shoe Spill in the North Pacific. *EOS Transactions of the American Geophysical Union*, vol. 73, n. 34, pp. 361, 365, 1992.

ocean currents until they made landfall along the western coast of North America. Knowing the location where the shoes entered the sea, and the eventual destinations they reached, enabled oceanographers to map ocean currents and better understand how the narrowing and broadening of the currents, and the eddies that they create, led to the transport and dispersal of a tracer, in this case the floating shoes. Shoes that entered the sea at essentially one point arrived along more than 500 miles of Canadian and American coastline. Such an experiment might have been envisioned and carried out by scientists, but in fact the entire episode was inadvertent. Actually, this experiment has been inadvertently repeated twice more, in January 1992 when containers with 29,000 bathtub toys washed overboard in the North Pacific, sending blue turtles, yellow ducks, and green frogs to the beaches near Sitka, Alaska, and again in December 1994 when some 34,000 hockey gloves were 'lost at sea'.

I recently visited a friend in the hospital and was required to don a sterile robe and latex gloves before entering her room. The reason, I was told, was that there were antibiotic-resistant bacteria loose in the hospital, and my friend, already weakened by her cancer, was particularly vulnerable to these superpotent microbes. They had acquired the resistance to antibiotics through another inadvertent experiment: the widespread and careless prescribing of antibiotics for maladies easily treatable in other ways. By using antibiotics when they were not essential, doctors had inadvertently given the harmful bacteria a glimpse into the medicinal arsenal. The bacteria took advantage of this opportunity by evolving mechanisms to cope with the antibiotics and so moved ahead in the never-ending war between microbes and medicine. Analogous effects have been observed in veterinary medicine, where livestock no longer respond to antibiotics that earlier had proven extremely effective in countering certain diseases. Similarly, antimalarial prophylactics prescribed for travelers and soldiers going to tropical areas have gradually lost efficacy as the malaria parasites have developed resistance to the new weaponry.

THE OZONE HOLE

Another large-scale inadvertent experiment began in 1929 when the DuPont Corporation began to market a new non-toxic, inert household refrigerant to replace the less desirable refrigerants of the day, ammonia (NH_3) and sulfur dioxide (SO_2). These latter compounds, while decent refrigerants, were both flammable and/or toxic and posed a hazard when used within homes. The new DuPont refrigerant belonged to a family of chemicals known as the chlorofluorocarbons (CFCs for short), compounds of chlorine, fluorine and carbon. These compounds, in addition to their excellent refrigerant properties, had many other desirable characteristics: if they leaked out of the refrigerator they would not ignite or combust, they posed no health hazard, and they were not soluble in water.

After World War II, other uses were discovered for the CFCs. They turned out to be good propellants for products that could be effectively distributed as aerosols. After all, they did not react with or dissolve into the materials with which they were mixed, and they could be compressed under pressure to serve as a propellant when released. Quickly, they became the propellant of choice for hair sprays, bug sprays, air fresheners, spray paints and myriad other products delivered to a target as an aerosol. Soon the CFCs were also being used in rigid foam products such as Styrofoam™ and polystyrene used in packing and insulation. Their refrigerant properties found new uses outside the household refrigerator, particularly in building and automobile air-conditioners. The CFCs were truly a great product, serving a wide array of perceived needs. However, as they became widespread, an inadvertent side effect slowly became apparent: the CFC molecules had made their way upward into Earth's stratosphere, where they became a key player in the destruction of stratospheric ozone.

Ozone is a particular form of oxygen in which three atoms of oxygen are bound together in a single molecule. Ordinary oxygen makes up about 20% of Earth's atmosphere and contains only two oxygen atoms. Ozone is produced and destroyed by natural processes in the stratosphere, yielding an equilibrium concentration that over the long term is very steady. Although falling far short of even 1% of the atmosphere, ozone plays a very important environmental role: it filters out much of the incoming ultraviolet radiation from the Sun. This radiation has deleterious health effects, and so the ozone in the stratosphere, even at its low concentration, is a very beneficial ingredient.

Unfortunately, the very properties of the CFCs that made them attractive industrial products also contributed to the unanticipated ozone destruction in the Earth's stratosphere. Each depression of a spraycan button delivered the CFCs to the atmosphere, and each discarded refrigerator, each junked auto with air-conditioning eventually leaked CFCs to the atmosphere. Once in the atmosphere, the inert CFCs would not chemically combine with other atmospheric constituents, and they were not washed out by rainfall because they were not soluble in water. They accumulated in the atmosphere, got stirred up by wind and weather, and some eventually made their way to the stratosphere where the ozone resides. Once in the stratosphere, they began to upset the equilibrium that existed between the natural processes of ozone production and destruction.

In the stratosphere, ultraviolet radiation from the sun provides sufficient energy to separate chlorine from a CFC molecule. The freed chlorine, with a voracious appetite for oxygen, finds an ozone molecule to attack. The chlorine pulls off and captures one of the three oxygen atoms, thus demoting the O_3 ozone molecule to an ordinary O_2 molecule. However, the chlorine atom's hold on its newly captured oxygen is shortlived. Single roaming oxygen atoms, spotting an oxygen in an embrace with a chlorine, will pull the oxygen away and join with it to form another ordinary oxygen molecule. The chlorine, newly stripped of its oxygen, seeks and finds another ozone molecule to take apart. Thus a single chlorine atom becomes a repeat offender, preying on many molecules of ozone over time. Left unchecked, this process could quickly lead to catastrophic ozone depletion. However, chlorine can also be stably incorporated into other molecules and thereby removed from action in the ozone wars. The process of sequestering the chlorine is, unfortunately, extremely slow; a single chlorine atom returns to attack ozone tens of thousands of times before it is finally captured or purged from the stratosphere.

The destruction of ozone by chlorine is particularly apparent in the south polar region over Antarctica. There the temperature averages a frigid $-45\,°C$ ($-50\,°F$) on an annual basis, and it is even colder during the polar winter when the sun disappears entirely for six months. The extreme cold promotes the formation of tiny ice crystals in the stratosphere. When solar radiation returns after the long winter darkness, the surfaces of these crystals provide a place for reactions that, when energized by the returning sun, convert inert chlorine-bearing compounds into other highly reactive ozone-destroying species. The rapid depletion of stratospheric ozone follows, a rite of spring, so to speak, over Antarctica.

Since the early 1980s, the destruction of the ozone over Antarctica has increased each year, exposing ever-larger areas of the southern hemisphere to the ultraviolet radiation. Since 1987, CFC production has been virtually eliminated through an international treaty known as the Montreal Protocol, and industrial chemists have taken steps to find alternatives for the CFCs in their many important roles. However, the unanticipated effects of introducing the CFCs will linger into the second half of this century as the CFC load in the atmosphere gradually diminishes.

Recognition that the CFCs were causing the destruction of the stratospheric ozone over Antarctica was the outcome of a scientific detective story that ultimately led to Nobel Prizes for the scientists who discovered the important processes involved. But it was a story that met many doubters along the way and initially inspired a great deal of resistance. Part of the resistance arose because of the large conceptual distance between cause and effect. The idea that ordinary human activity, such as the use of hair spray, could have a significant effect on Earth's atmosphere is an alien concept. We have grown up with the image of humans as puny inhabitants of a big planet that displays natural forces much more powerful than anything we might contribute. We have evolved to make us alert to things we can see, hear, touch, and smell, things in our immediate environment that might be a threat. In the twentieth century, it has been an intellectual stretch to realize that the slow accumulation of effects from human activity can, in the aggregate, affect the entire Earth in significant ways that also constitute a threat to us.

The inadvertent experiment of the CFCs and ozone depletion had the salutary effect of a wake-up call, in that it made scientists and many others aware of the collective impact that humans have on the global environment. The scientific studies that discovered the depletion and identified the causes, and the political response that ultimately led to international steps toward remediation, have somewhat prepared the world to confront an even larger inadvertent experiment, global climate change.

CHANGING CLIMATE

Perhaps the greatest inadvertent experiment in human history, that of global climate change, is currently underway. The focus of interest in this experiment is how the global climate system is responding to ever-increasing levels of carbon dioxide in the atmosphere. In the list of atmospheric constituents, carbon dioxide ranks a distant fourth place, well behind nitrogen and oxygen, which together make up 99% of the atmosphere by volume. It even stands behind argon, which accounts for about nine-tenths of the remaining 1%. In 1750, just as the industrial revolution was getting underway, carbon dioxide accounted for only about 280 of every million units of atmospheric volume, or about 0.028%. Today, following two and a half centuries of burning fossil fuels to power the global industrial economy, the level of carbon dioxide in the atmosphere has grown to about 380 parts per million (ppm). And, under virtually every reasonable scenario of future global population, energy usage, and technological changes, carbon dioxide will grow to around 500–600 ppm early in the second half of the twenty-first century, a concentration about twice the pre-industrial level. This is a level greater than any seen in the geological record for the past half-million years.

Do we need to be concerned about changes in an atmospheric constituent that is so small, a fraction of the total volume of the atmosphere? The answer is clearly yes. Small or not, carbon dioxide plays a vitally important role in regulating Earth's surface temperature and climate, through its action as a greenhouse gas. The term greenhouse refers to the ability of certain atmospheric gases to trap energy that Earth is radiating away from its surface; carbon dioxide, methane, water vapor,[8] the CFCs, and other gaseous constituents present in the atmosphere in trace amounts exhibit this trapping property. And we should be very happy that these greenhouse gases are present, because without them, Earth would have a surface temperature well below the freezing temperature of water, and our home planet would be an icehouse. That we have oceans, lakes and rivers on Earth, instead of a cover of glaciers, icecaps and sea ice, is a consequence of the presence of a greenhouse atmosphere. Moreover, we know Earth has had this greenhouse since its earliest days, because there is ample evidence throughout the geological record for sedimentary rocks deposited in water. The presence of water, instead of ice, throughout most of Earth history, requires a greenhouse blanket of long standing.[9]

In recent centuries, through the combustion of coal, petroleum, and natural gas, the addition of carbon dioxide to the atmosphere is changing Earth's natural greenhouse, making it more effective as a blanket covering the planet. How this warming of Earth will play out in the coming decades, in terms of changing sea levels, diminishing ice, shifting precipitation patterns, extreme weather events, and altered vegetation and agriculture, is not entirely clear. But many changes are already underway, and further change is unavoidable, even if remedial steps were to be initiated soon. Whether Earth's climate

---

[8] The concentration of water vapor in the atmosphere is highly variable regionally, depending on the temperature of the atmosphere.
[9] Water vapor may have played an important role in setting Earth's surface temperature very early in Earth history.

will change slowly and incrementally, or suddenly and catastrophically, is a major uncertainty with significant implications.

The reaction of the public to the protracted and heated debate over this 'natural' experiment was initially one of disbelief, for the same reason they were initially skeptical about the ozone depletion: many found it hard to imagine that each individual's day-to-day activities – heating and lighting their homes and driving their cars – could make a difference in large-scale and powerful natural processes. Add to that the deliberate and self-serving public relations campaigns by the affected industries, assuring the public that the science was immature and unpersuasive.

What has been accepted by scientist and skeptics alike is that carbon dioxide has been increasing year by year to levels not seen on Earth for a long, long time. And no longer is there much scientific debate about whether Earth is warming, or that human-produced carbon dioxide is playing a substantial role. The discussion now centers on what the outcome of this experiment will be, what consequences we should anticipate, and what we should be doing to counteract or accommodate to an altered climate. I will return to this great experiment in some detail in the final chapter of this book.

Long before humans appeared on Earth, nature had already run many such experiments, none of which we can repeat, but the results of which we can observe. Continents have drifted from tropical latitudes to polar regions, former ocean basins have been elevated into mountain ranges, and changing ocean circulation has altered the global climate. Life on Earth has had many turning points, with both emergence and extinction of species resulting from environmental changes brought about by geologic processes. We can only try to understand these natural experiments by studying the outcome; we cannot run the experiments again, nor do we have the capacity to alter conditions of the experiment.

Reconstructing the past, in both a historical and a geologic sense, involves posing questions and seeking answers, an experiment of sorts.

But it is not an experiment we design or control; it is an experiment that has been carried out in our absence, and we only get to see the outcome. We want to understand the processes and circumstances that have yielded what we see in the present day, but we will encounter uncertainty as to whether any scenario we reconstruct is actually what happened. In the next chapter we will delve into some of the special types of uncertainty that accompany reconstruction of the past, and see how they can be accommodated.

9   Reconstructing the past

The farther backward you can look, the farther forward you are likely to see.

Winston Churchill

The historical sciences such as archeology, geology, and astronomy are burdened with a special form of uncertainty known as non-uniqueness. When trying to understand why something happened the way it did, these scientists must try to reconstruct the circumstances of the event, and make hypotheses about the processes that governed the event. But as we try to reconstruct an historical event, we deal with an incomplete record. And with an incomplete record of an only partially understood process, there may be more than one way that the evidence can be explained. In other words, we must live in the shadow of non-uniqueness. At any given time, the incomplete evidence may admit many interpretations, and at a later time, additional evidence may eliminate some of those possibilities.

Dealing with uncertainty about the past is a way of life with geologists, who in their work of reconstructing natural history are always working with half a deck or less. Nature is not a mindful conservator, and the inevitable consequence of time is that the record of what happened long ago becomes degraded and fragmentary. In their efforts to understand and interpret incomplete information, geologists always work with a handful of provisional scenarios relevant to explaining their observations. This mode of thinking was enshrined by Thomas C. Chamberlin, a prominent geologist and President of the University of Wisconsin in the late nineteenth and early twentieth centuries, and later President of the American Association for the Advancement of Science. In 1890, Chamberlin published an essay[1] titled *The Method of Multiple Working Hypotheses*, in which he laid out the

[1] *Science*, 7, February 1890.

philosophy of not placing all of one's interpretive eggs into a single basket. Rather, he argued that, when one ran up a blind alley, the sooner one backed up and chose another avenue of pursuit the better. He argued that having many alternatives in hand promoted critical thinking and prevented scientists from developing mental ruts. Chamberlin's concept is echoed in the wisdom of an anonymous savant: "Nothing is more dangerous than an idea when it's the only one we have".

A PLANE IS DOWN ...

Several aviation tragedies in recent years have reminded the public that the hazards of flying, though far smaller than the risk of dying in an automobile accident, are not infinitesimal. The crash of TWA Flight 800 in 1996, of the Air France Concorde in 2000, and American Airlines Flight 587 in 2001 – all have been extensively investigated, and all continue to have varying degrees of uncertainty in reconstructing and understanding the events leading to their respective crashes.

TWA 800, because it occurred earlier and has been investigated longer than the others, provides a useful example for a discussion of the uncertainties of reconstructing past events. Shortly after its departure from New York's Kennedy International Airport on 17 July 1996 en route to Paris, the aircraft exploded in mid-air over Long Island and plunged into the Atlantic Ocean. Everyone aboard perished. Various scenarios were advanced to explain the event, and evidence, albeit incomplete, was gathered from the sea and elsewhere to evaluate and assess the likelihood of each scenario. The entire event was cloaked in great uncertainty about what had happened.

With the 1988 bombing of Pan Am flight 103 over Lockerbie, Scotland still burning in the international memory, it was no surprise that the early scenarios included a wide range of hypotheses about acts of terrorism: a bomb placed aboard the aircraft in the baggage hold, a bomb in the passenger compartment that evaded detection in the security screening, a bomb smuggled aboard with the complicity of a security operator, a bomb brought aboard in the meal service containers, a bomb hidden on board by a member of the cleaning crew, a terrorist attack with a ground-to-air missile.

A second broad category of hypotheses focused on mechanical malfunctions of the aircraft: an explosive decompression of the aircraft, an internal fire in one of the galleys, an engine fire that exploded the fuel tanks, an explosion in the fuel tanks from static electricity or a short circuit. Yet a third category of possibilities included some improbable but minutely possible causes: a lightning strike, a collision with a small aircraft, an encounter with a flock of high-flying geese, or severe clear air turbulence that dismembered the aircraft.

The investigation proceeded on many tracks. The aircraft was almost fully reconstructed from debris retrieved from the ocean bottom. The radar tracking of the ill-fated flight was scrutinized. Eyewitnesses on the ground and pilots of other nearby aircraft gave testimony. The security gate operators, baggage handlers, catering service providers, and cleaners were interrogated. Experiments were conducted with similar fuel tanks, to determine under what conditions an explosion might ensue. Four years later, following the longest and most expensive crash investigation in its thirty-three year history, the US National Transportation Safety Board issued a draft report[2] on the possible causes of the demise of TWA 800. After exhaustive analyses and evaluations of the many scenarios, the NTSB rejected the terrorist hypotheses in their many variations. Similarly, one by one, it dismissed the improbable events.

Potential mechanical malfunctions became the central focus of the investigation. Eventually strong evidence pointed to an explosion in the center wing fuel tank, which, ironically, was nearly empty, save for a volatile mix of fuel vapors and atmospheric oxygen. The mix was particularly unstable and vulnerable to ignition because, prior to takeoff, the aircraft had sat on the tarmac for some three hours under a hot summer sun, raising the temperature inside the fuel tank. However, as to what actually ignited the explosion in the empty fuel

[2]Dated 22 August 2000; www.ntsb.gov

tank, there is still no definitive, uniformly accepted answer. Frayed insulation in a fuel guage circuit is the prime suspect.

Uncertainty remains in our understanding of the TWA 800 tragedy, and always will. Some 10% of the airplane remains on the ocean floor, lost forever. We cannot interview the flight crew to learn what they experienced in the fateful seconds following the explosion. We cannot re-enact the flight under controlled experimental conditions to test some of the ideas about the cause of the explosion. Yet, even with this admittedly incomplete array of evidence, the range of possible explanations has been narrowed considerably, and a few explanations now have a much higher probability of being true than the many others that the NTSB rejected in its exhaustive investigation. Nevertheless, conspiracy theories still float widely on the Internet, including allegations of a government cover-up of international terrorism. The proponents of these theories are in many ways examples of 'true believers', people who have their minds made up no matter what the evidence is. No facts or logical reasoning will budge them from their beliefs. Because of their inability or unwillingness to accept evidence that renders their pet beliefs untenable, we must be able to set aside the uncertainty that they introduce into the discussion.

Uncertainty is not without its benefits. Because we do not know exactly the cause of the fuel tank explosion, we are forced to entertain several possibilities, and take remedial steps to address this multiplicity. Were the fuel tanks too warm? Perhaps the aircraft manufacturers could install better insulation to shield the tanks from the heat of the tarmac. Was the mix of fuel vapor and atmosphere too volatile? Perhaps inert nitrogen could be pumped into the tanks as fuel is consumed, to reduce the volatility. Was the insulation on the fuel gauge circuit too thin? A thicker jacket might help. The investigation of the Concorde crash showed that its fuel tanks were too vulnerable to puncture by debris flying off blown-out tires, and the failure of the tailfin on American Airlines 587 has led to attention being focused on the durability of carbon fiber composites. The investigations into the reasons behind each of these tragedies were shrouded in uncertainty at the outset. But in each case the investigations revealed a previously unsuspected design flaw and have led to improvements in design and materials. In short, the initial uncertainty spurred creative thinking; ultimately, commercial aircraft will become even safer than they already are.

In what ways is the saga of TWA 800 instructive in helping us to understand scientific uncertainty, particularly in reconstructing the past? Historians, geologists, and archeologists routinely must address events that have taken place long ago, and they always must deal with an incomplete body of evidence. Which is not to say that no progress has been made in understanding human history, or the origin and the subsequent evolution of Earth. To the contrary, we have learned a great deal and will continue to learn more as additional evidence accumulates and concepts undergo revision. But even as some scenarios are set aside, some uncertainty about what actually happened will remain.

IN THE COURTROOM

Another venue where incomplete, inaccurate, and conflicting evidence is the norm is in the courtroom. In the halls of justice, reconstructing the past is an everyday occurrence. Whodunnit? That question and its myriad variants are addressed through interrogation, recollection, speculation, forensic or physical evidence, psychology, and more. Every example of litigation involves competing versions of what has happened (in science, as mentioned earlier in this chapter, these are called multiple working hypotheses).

Incomplete evidence? As we learned in the well-known case of Timothy McVeigh, the perpetrator of the 1995 bombing of the Federal Center in Oklahoma City, the FBI inadvertently did not present thousands of pages of evidentiary documents to the defense prior to the trial. In more ordinary settings, the codicils to a will may have disappeared, the premarital contract presented to the court included a list of only her assets, or the murder weapon and getaway car were never found.

Inaccurate evidence? A ballistics test on a bullet from a suspected murder weapon is inadvertently labeled to be from a different handgun. A witness testified that the license plate number of the getaway car was COR-134, when in fact it was C0P-1B9. And as time elapses between the crime and the trial, memories become foggy, witnesses may die, and evidence degrades.

Conflicting evidence? Prosecution witnesses testify that at the time of a crime the defendant was at the crime scene, while defense witnesses testify that the defendant was with them, 2000 miles away. A fingerprint at the scene of the crime is that of the defendant, but a bloodstain on the murder weapon belongs to neither the victim nor the defendant. The president of a company involved in a product liability suit, according to some internal documents was aware – and according to other documents unaware – of a safety hazard in one of the company's products.

Judges and juries must often sift through competing scenarios, incomplete and erroneous evidence, and outright conflicting and often mendacious testimony. They listen carefully, weigh and evaluate the evidence, and reach decisions. Jurors are not permitted the option of seeking more evidence, or engaging in lengthy research that might perhaps clarify an important but murky point. They cannot postpone making a decision to await a possible reduction of the uncertainty surrounding their case. They are asked to reach verdicts that are 'beyond a reasonable doubt', not 100% certain.

Do juries ever make mistakes? We know that they do. But we also know that convictions are sometimes reversed on the basis of new physical evidence, recanting of testimony, or subsequent confessions from others. We accept imperfections in the legal system, for the greater benefits we attach to our constitutional right to speedy judgment by a jury of our peers. We recognize that the courtroom is an untidy arena where decisions, usually right but occasionally wrong, are made in the face of incomplete, inaccurate, and conflicting evidence.

SUBMERGED

An odd but interesting validation of the benefits of being able to cope with incomplete, inaccurate, and sometimes conflicting information emerged from a post World War II study of the qualities that characterized successful submarine captains. In this study, the definition of a successful captain was simple: those that survived the war were successful, whereas those that died were not. The study looked into, among many factors, what educational background the captain brought to the assignment. The results of this study concluded that people with training in fields such as geology and economics were more frequently survivors, whereas mathematicians and theoretical physicists were more prone to being killed in battle.

The interpretation offered for this outcome was that, submerged and under attack, submariners worked in a hostile environment marked by incomplete, inaccurate, and conflicting information. Acoustic listening devices provided some information about their enemies' positions and activities, and exploding depth charges, when not too close by, gave some insight into the attack plan of their adversaries. Geologists and economists, themselves scientists accustomed to working with fragmentary empirical data, were readily able to formulate reasonably accurate scenarios of what was happening and take evasive action. Mathematicians and physicists, by comparison, accustomed to highly structured scientific thinking according to well-defined axioms and rules, were less able to deal with observations that did not fit easily within such a rigid template. Or worse yet, with observations that seemed to conflict with each other. Submarine warfare was a very untidy situation, one that often fell outside the structured logic that mathematicians and physicists were accustomed to. While they fretted about the imperfections of the situation, the blemishes or holes in the fabric of the logic, they were felled by a depth charge.

The fatal flaw for the mathematicians and physicists was their intuitive dislike of problems that did not fit well into the theoretical frameworks they were familiar with, and of fragmentary information that left open too many options for them to evaluate. They were accustomed to solving problems that had a right answer, not many possible answers. Less-mature sciences, such as geology and economics, had many fewer such constraints, and therefore captains with such backgrounds proceeded more intuitively and less analytically. And, so the story goes, they were more likely to have survived.

$2+2=4$, BUT WHAT DOES '$?=4$' MEAN?

Numerical models can help us to understand the past by narrowing the range of possible pathways to an historical outcome. In using computers to study how and why things turned out the way they did, scientists sometimes employ a type of mathematics known as *inversion.* Inverse models let us undertake reconstruction on a quantitative basis. Without delving deeply into the mathematical underpinnings of inversion, one can easily get a feeling for the logic of this approach. Consider the familiar problem, $2+2=$?. In this problem we are given two quantities (2 and 2), and a rule for combining them (+); with the usual meaning attached to all of the symbols, most people reach the well-accepted answer of 4. When both the quantities and the rules by which we manipulate them are straightforward and unambiguous, then the outcome is unique. We call this type of calculation a 'forward' model, one in which we provide ingredients and a recipe and obtain a unique outcome.

Now let us consider a variant of this problem, stated simply as $?=4$. This is an *inverse* problem, a reconstruction problem, where we are given a result, 4, and are asked to determine what led to the result. Immediately one might protest "That's unfair, there are lots of ways to produce 4! There is no 'correct answer'." Indeed, there are many ways to produce four: $5-1$, $3+1$, $8\div2$, 2×2, the square root of 16, and of course our old favorite $2+2$. And we need not restrict our options just to adding integers. How about $3.47+2.85-2.32$? You get the idea. Mathematicians will tell us that in fact there exists an infinity of ways to produce this given result, that there are many, infinitely many, scenarios that adequately account for the existence of this given outcome. So what do we do? Just throw up our hands in futility? Moan that the problem is insoluble and just give up? No, we try to bring other information to bear that will help us to narrow the range of possibilities.

In terms of a real-world example, imagine a box sitting on the table that we are told contains $4. We are asked to determine what combination of bills and coins the box contains. This is an example of the $? = 4$ problem. How can we narrow the number of possibilities? A quick shake of the box reveals no coins rattling around; therefore, we are dealing only with paper notes. A little careful thinking about the problem enables us to conclude that we are dealing only with the process of addition, that is, we are being asked what combination of bills *add up* to $4? Next we recall that the only banknotes with face value less than $4 are the one and two dollar bills. Now we can make some progress. There are only three possibilities: two $2 bills, four $1 bills, or one $2 bill and two $1 bills. The infinitely many answers to the general problem of $? = 4$ have been reduced to only three by incorporating additional knowledge about the specific problem at hand. One could speculate even further by using information obtained from the US Treasury about the relative abundance of $1 bills versus $2 bills in circulation, and thus place probability estimates on each of the three possible solutions. Uncertainty will remain, but it has been considerably reduced.

Another approach to finding a family of acceptable solutions to a problem is a brute force method involving trying out many provisional answers and checking each one to see if it qualifies as a possible answer. Let us examine, for example, the problem $? = 25$. This, like our earlier example, has an infinite number of solutions. But if we put some additional constraints on the procedure we can narrow the range of possible solutions. Let us restrict ourselves to sums of two positive integers (whole numbers, not fractions), and then program a computer to test many different combinations and to keep a list of those pairs of numbers that add up to 25. In so doing, we have actually asked the computer to construct a large number of forward models and test each one for its suitability as a solution to our problem.

Computers seldom complain about this kind of grunt work, but we can help the computer out a little in the following ways: obviously the numbers must each be less than 25, or their sum could not possibly be 25. So we tell the computer to restrict the numbers from which it draws to those between 1 and 24. And, we know that 25 is an odd number, and therefore two even numbers, or two odd numbers can never add up to 25. So when the computer randomly generates two numbers to add together, we can tell it not to bother adding them together unless one is even and one is odd. Then we turn the computer loose, with instructions to repeat this operation 5000 times.

Each time the computer finds a pair of numbers that work, it adds them to the list of possible solutions. But because each trial is conducted with no knowledge of any previous trial, it is entirely possible that in 5000 trials it might find the same pair of numbers more than once. So before adding this pair to the list again, we ask the computer to check and see if that pair is already on the list. If so, keep a count of the number of times that the repetition occurred, and then move on to try out another pair of numbers. At the end of 5000 trials, the computer prints out a list something like this:

1 + 24 (4 times)	7 + 18 (5 times)
2 + 23 (7 times)	8 + 17 (4 times)
3 + 22 (3 times)	9 + 16 (3 times)
4 + 21 (6 times)	10 + 15 (6 times)
5 + 20 (5 times)	11 + 14 (4 times)
6 + 19 (4 times)	12 + 13 (2 times)

In 5000 trials, the computer discovered 12 combinations of integers that summed to 25. Moreover, it found each one of them several times, giving some confidence that it is unlikely that it missed any combinations, although when selecting numbers randomly it would certainly be possible to miss one or more of these solutions to our problem. Had we drawn numbers only 100 times, it is likely that we would have missed some satisfactory combinations altogether. The greater the number of trials, the higher the probability that you will find more, or even all, of the possible solutions. But in a real problem, you will never know for sure that you have not overlooked some of the solutions. The additional information that we employed narrowed the search considerably and helped to zoom in on the range and type of numbers where we had reason to think solutions might be present.

In the broader context of real-world problems, the choice of where to explore for solutions is a very important part of finding them. Just like exploring for gold, if you know something about the geological setting of previous discoveries of gold, your chances of finding a new deposit are greatly enhanced, at least compared with searching on a totally random basis.

This brute force approach to finding a collection of outcomes that satisfy all of the conditions of a problem is called _Monte Carlo inversion_, because it relies on the laws of probability to ensure that if you try enough candidate solutions in a well-designed search you have a decent likelihood of finding many that meet all the requirements. Actually, the Monte Carlo approach is just a repetitive application of the forward model. In the aggregate, the method identifies both possible and impossible solutions to a problem, and for the possibles it develops some probabilistic estimates of the likelihood of each one. Real scientific problems, dramatically more complex than the simple example we just examined, sometimes test tens of millions of candidate solutions to find those relatively few that meet every condition we impose.

One example of a Monte Carlo analysis relates to interpreting a recently observed temperature profile down a deep borehole drilled into the Greenland ice sheet.[3] Geologists measure the temperature within the Earth for a number of reasons. One is to determine how much heat is flowing to the Earth's surface from the deeper interior. This is a quantity that varies from place to place and is related to the

[3]Dahl-Jensen, D., Mosegaard, K., Gundestrup, N., et al., Past temperatures directly from the Greenland ice sheet. _Science_ 282, vol. pp. 268–271, 1998.

tectonic stability of the terrain. A second reason is to reconstruct the temperature history at the surface and thus gain a glimpse of Earth's fluctuating climate backward in time. This latter endeavor is possible because as the climate changes at the surface, the material beneath the surface will feel the change. Consequently, a prolonged cooling over several centuries will affect rock temperatures (or ice temperatures as in the case of the Greenland borehole) to a depth of about a thousand feet. An ice age will chill the rocks to depths of more than three thousand feet.

The temperature profile in any borehole is a composite of both the deeper heat flowing upward and the climatic fluctuations traveling downward. A Monte Carlo analysis of the temperatures measured in the ice borehole in Greenland[4] tried out over three million combinations of the deep heat flow and surface climate histories and retained just 2000 of these trials as adequate explanations of the temperatures actually measured. Most of these 2000 solutions display very common characteristics, and while they do not tell you exactly what has happened with 100% certainty, they do define a rather small range in which the actual climate history experienced at this site in Greenland very likely sits.

BAYES AND BOREHOLES

Frequently in the world of scientific observation, scientists are faced with observations and explanatory hypotheses that just do not mesh. It is like trying to put a square peg into a round hole. The two just do not fit together. But is it the peg or the hole that is at fault? What adjustments to each might make the two compatible? A larger hole or smaller peg might be one solution. A not-quite-round hole, and not-quite-square peg, might also enable a fit. One type of inversion, called a _Bayesian_ analysis after Thomas Bayes, an eighteenth century British clergyman with a bent for mathematical analysis, gives a quantitative

---

[4]In this context ice behaves just like rock, or any other solid for that matter, in its ability to conduct heat.

estimate of the probability of various scenarios, given the level of confidence one has about all the factors in the case.[5] If we have reason to believe that the shape of the peg is less well known than the shape of the hole, a Bayesian analysis will estimate the most probable shapes of both, consistent with our prior knowledge about both peg and hole, and our hunches about why they do not fit. This Bayesian approach is conceptually quite different from the more traditional methods of estimating probabilities, because it allows the scientist to exercise judgments that express knowledge and confidence about all the factors in the problem.

The more traditional approach to estimating probability is called the *frequentist* approach, which draws inferences from the frequency with which an event occurs. In the classic coin-flip problem, where we ask what is the probability that on any given flip we will observe heads or tails, the frequentist would flip the coin many times, tabulate the number of heads and tails that come up, and calculate a probability from the frequency that heads and tails appeared. A Bayesian analyst, however, might approach the problem quite differently. It is clear that there are only two possible outcomes, and by invoking prior knowledge of the physics of tumbling airborne metallic discs, an experimenter could determine that there is no physical reason that the coin should preferentially fall face down or up. This prior knowledge would suggest an equal probability for heads or tails, without need of multiple coin flips to establish that experimentally.

We have already discussed an example of a frequentist analysis when we examined polling results prior to an election. The statement that there is a 95% probability that candidate Smith's vote total will fall in the 38–46% range means that if the poll were repeated one hundred times, candidate Smith's total would be in that range ninety-five times. A frequentist views probability as a frequency of occurrence and evaluates a hypothesis in terms of the likelihood of

[5]A short and accessible description of Bayesian methods is given by David Malakoff in *Science*, vol. 286, pp. 1460–1464, 1999.

getting the same result if one repeats an experiment many times. A Bayesian analyst does not think about repeating an experiment but rather asks the question "How many hypotheses is this single set of observations consistent with?" As in the case of the peg and the hole, it will determine what adjustments to both the peg and the hole could make the experiment and its interpretation self-consistent.

In my own work in geophysics, I use a Bayesian analysis to make inferences about recent climate change experienced at the Earth's surface. As I mentioned briefly in the notes about the author, my research centers on taking the Earth's temperature. This measurement is accomplished in boreholes that penetrate the rocks of the Earth to depths of several hundred meters below the surface. We lower a sensitive thermometer down the hole, pausing to read the temperature every ten meters or so. The result is a series of measurements that in the aggregate yield a profile of the temperature down the borehole. In the discussion above about Monte Carlo methods of analysis, I mentioned one of these temperature profiles taken in a borehole in Greenland. My international colleagues and I have collected similar profiles from more than seven hundred boreholes on all of the continents, and we have analyzed these profiles to reconstruct the surface temperature history experienced at each of these sites.

In this context, a Bayesian analysis proceeds as follows: we have a collection of temperature measurements taken at various depths, and often we also have samples of the rock from the length of the borehole, obtained during drilling. We are able to measure the heat transfer properties of the rocks, to determine how well and how fast they conduct heat. This physical property plays a significant role in estimating how long it takes a warming at the surface to travel downward and affect the rock temperature at various depths. Recall, of course, that the measurements of temperature and rock properties have a range of uncertainty related to the thermometers and other instruments used to measure the rocks' heat transfer capability.

Next we make an educated guess (i.e. propose an initial model) as to what surface temperature history might be responsible for the subsurface temperatures we observed, given what we know about the heat transfer properties of the rock. This initial model is a hypothesis to be tested for its compatibility with the observations. The compatibility test requires a computation of what the consequences of the initial guess would be for the subsurface temperatures, and a comparison of those calculated temperatures with the observed temperatures. Never will the observed and computed temperatures be exactly the same, and so the computer must now address why the mismatch, small or large, occurs. Is it because of the uncertainties in the temperature or depth measurements down the borehole or in the determination of the properties of the rock? Is it because the rock has been sampled inadequately, and we have missed some important changes in the heat transfer properties of the rocks? Is it because our initial guess for the temperature history was way off the mark? Is it because the heat transfer theory we used to calculate the propagation of surface temperature changes downward into the subsurface is inadequate or incomplete?

A Bayesian inversion will take all of these factors into account and make its best estimate of how much of the misfit is the result of measurement errors, how much is from insufficient sampling, how much from an incomplete picture of the heat transfer mechanisms, and how much from a poor guess of the history. It will then suggest how each of these quantities must change in order to be consistent with each other. It is up to the human interpreter to tell the computer how much leeway it can have in making the adjustments. And it is in this latter endeavor, estimating how much give and take there may be in the measurements, in the theory, and in the initial estimate of the surface temperature history, where scientific judgment and experience play a crucial role.

There are two extreme cases that usually can be dismissed as overly rigid. The first is one in which the scientist insists that his or her initial guess about the history is absolutely correct, and if the subsurface temperatures do not agree with it, then there is something wrong with the temperature measurements. This is an example of the rigidity I mentioned earlier. "My mind is made up – if the observations don't agree with my model, then look for errors in the measurements to explain the disagreement." A second rigidity is in the opposite direction. "The observations are absolutely correct, and if the model calculations do not agree with these observations, then the model or the theory from which the model calculations emanated needs to be revised."

In practice, of course, we know there are uncertainties in the measurements, and that the heat transfer theory we employ is only an approximate description of how heat moves through rocks. And we certainly do not believe that we know the surface temperature history perfectly in advance! So we assign a range of flexibility in which the computer can make adjustments for each component of this problem, and ask it to tell us how best to fit every piece of this puzzle together.

In our interpretations of the rock temperatures, we make a very conservative initial guess of the climate history, one that asserts that there has been no climate change at all. This is called a 'null hypothesis'. Along with presenting the null hypothesis as a first guess, we also tell the computer that we are willing to deviate from that conservative hypothesis if the temperature observations push in that direction, but we impose limits on how big an adjustment may take place. Next the computer interrogates the subsurface temperatures to see if they are consistent with this hypothesis within the assigned range of uncertainty of the temperature measurements. If consistency cannot be achieved within that range, then the computer turns to adjusting the null hypothesis of the climate history to one that is more compatible with the subsurface temperature.

At the end of the procedure, we see the Bayesian estimates of how the model could be revised to be consistent with the temperatures, simultaneously with how the temperatures might be revised to be consistent with the revised model. It is a highly fluid procedure, where every part of the problem, the observations, the theory, and the interpretation, is fair game for some adjustment. The range of permissible adjustments are governed by our best judgment of how well we know each component, and our best estimates of which element can bend the most or least. If the computer reaches an impasse, where it cannot make a coherent story out of the *pot pourri* within the constraints we have placed on each, it beeps in protest and says it cannot help us any more.

Uncertainty is mingled with every ingredient of this complex recipe, and it permeates the cooking instructions as well. But the Bayesian inversion procedure parcels the uncertainty out unevenly, and in the end we understand the robustness of our observations, our theory, and our 'answer', the reconstructed history, much better. If we carry the cooking analogy one step further, it is entirely possible, that, when we leave the kitchen, if the cake has fallen or is burnt, we will have not only an imperfect product but also estimates of the probability that the ingredients were bad, the recipe was faulty, the stove malfunctioned, or the cook did not follow directions.

What have we learned about climate change from this experience with more than 700 borehole temperature profiles from around the world? They tell a remarkable story, one that is independent of the surface temperature measurements made in meteorological observatories or by floating buoys and ships at sea measuring sea surface temperatures. The underground temperatures show that the rocks have been warming by 1 °C (almost 2 °F) over the past five centuries. But fully half of the warming has taken place in the twentieth century alone, and an additional 30% in the nineteenth century. This warming as seen in the rocks is fully consistent with the record of warming reconstructed from the surface temperature observations, thus confirming with data from the subsurface what the historical surface observations have revealed about global warming.

CORRELATION AND CAUSATION

In reconstructing the past we want to know not only *what* happened but, if possible, to also learn *why* it happened. For example, we know that dinosaurs became extinct at the end of the Mesozoic Era, but we would also like to know what factors led to the demise. To help to answer questions of causation, scientists will often try to see if some factor is behaving in a pattern which suggests that it might be related to the event they are trying to understand. In the field of climate change, it is now well established, and even well accepted, that Earth's average temperature has a clear warming trend over the past century. But what is causing the climate to change? Is it increased radiative output from the sun? Is it because of the addition of greenhouse gases in the atmosphere from the burning of fossil fuels? Is it because of volcanic activity? Or possibly is it caused by some combination of these effects?

One approach that climate scientists use to address these questions is called *correlation analysis.* By reconstructing the history of greenhouse gas concentrations in the atmosphere over the same time interval as the temperature reconstruction, we can see that the upward trend of temperature in the twentieth century has been accompanied by an upward trend in greenhouse gas concentrations in the atmosphere. The behavior of each appears to have a similar pattern over time, and we describe this similarity by saying that the two phenomena, Earth's mean surface temperature and the greenhouse gas concentrations in the atmosphere, are positively correlated. The correlation can be placed on a quantitative basis through the calculation of something called a correlation coefficient.

Without examining the intricacies of this calculation, let me just note some of the properties of the correlation coefficient: if every small change in Earth's surface temperature was paralleled by a proportional change in the greenhouse gas concentrations, ups with ups and downs with downs, we would calculate a correlation coefficient of $+1.0$ and call the two perfectly correlated. If the ups of one were accompanied by downs in the other, and vice versa, we would calculate a correlation of -1.0 and call the two perfectly anticorrelated. If the ups and downs of one appeared to be totally unrelated to the ups and downs of the other, we would calculate a correlation coefficient near zero, and we would say the two were uncorrelated and probably not related.

Correlation, however, is not the same as causation. Things can be well correlated, suggesting they may be related by some process, but the process itself is not identified. A classic example, a favorite of every statistics teacher, is the story of a person fascinated with fires, who raced to every location of a reported fire and duly noted that fire trucks from the municipal fire department were present. His logbook of conflagration, day by day, year by year, confirmed a strong correlation of the occurrence of fire and the presence of fire trucks. Can one dismiss the inference from this set of observations that fire trucks cause fires? The logbook alone documents only the correlation; more careful observations about which came first, the fires or the fire trucks, would be necessary to ascertain which was cause and which was effect.

Another well-known correlation is the high occurrence of malaria in countries with low per-capita gross domestic product.[6] Reduced to simplicity, malaria and poverty march hand in hand in many places. But unlike the fictitious example of the correlation in the geography of fires and fire trucks, where determining which came first might lead to some understanding of cause and effect, malaria and poverty feed on each other. Each is a cause, each is an effect. Malaria surely impedes economic development through its effects on fertility, individual productivity, absenteeism, health care costs, and individual and national accumulation of capital. But a weak economy just as surely promotes malaria by not providing a civil and public health infrastructure that would eliminate mosquito breeding areas, provide medical prophylaxis and treatment, and education. What we can learn from this correlation of malaria with poverty is that a curtailment in one will likely also lead to a reduction in the other. Or sadly, an increase in one will likely lead to an increase in the other.

The question about the role of greenhouse gases in climate change has been similarly studied. The longest records of temperature

[6]Sachs, J. and Malaney, P., The economic and social burden of malaria, *Nature*. vol. 415, p. 680, 2002.

and greenhouse gas concentrations in the atmosphere come from analyses of a long ice core in Antarctica. At the high elevation and polar setting of Antarctica, snow accumulates annually, with each year's snowfall buried by the deposit that comes in the following year. The fluffy snow is over the next few years gradually compressed into glacial ice, and some of the air trapped in the snow is sequestered into small bubbles trapped in the ice. The air in the bubble is a sample of the atmosphere in the interval between snowfall and recrystallization into ice. It can be analyzed to reveal how abundant the greenhouse gases were in the atmosphere prior to being sealed off and preserved.

Drilling on the Antarctic polar plateau has penetrated almost four kilometers (more than 12,000 feet) of snow and ice that has accumulated over the past 420,000 years. The temperature at which the snow precipitated can be determined by the ratio of hydrogen isotopes (heavy hydrogen versus light hydrogen) in the water molecules of each ice layer. Chemical analyses of the air trapped in bubbles have shown that changes of temperature and changes of greenhouse gas concentrations in the atmosphere are very well correlated: when greenhouse concentrations were high, so was the temperature, and similarly when one was low so was the other. But that correlation alone does not tell us that greenhouse gas changes cause temperature changes, or whether temperature changes cause greenhouse gas changes. Just as with the fires and fire trucks, a detailed chronology of the changes might prove helpful. Did one occur before the other, or vice versa? In the case of the almost half-million year history contained in the Antarctic ice cores, the temporal resolving power diminishes as one attempts to look further back in time, and the question cannot yet be resolved unequivocally. Perhaps the temperature increased first and was closely followed by the greenhouse gases, or perhaps it was the other way around. Either interpretation requires that there be a mechanism that couples the two together.

What might that coupling mechanism be? Almost certainly it is a third player in the global climate system, life on Earth. Life enters the equation because it is part of the carbon cycle on Earth. Carbon is a constituent of all life forms, plant or animal, large or small, on land or in the oceans. Carbon moves into the atmosphere from former living systems, as plants die and decay or through the combustion of carbon-based fossil fuels such as coal, oil, and natural gas. Combustion and decay both involve carbon combining chemically with oxygen to yield the gas carbon dioxide. Some CO_2 is absorbed by the oceans as atmospheric concentrations grow.

Climate, carbon, and life are intricately intertwined and always have been. A change in one will lead to a chain reaction of changes in the others. What is really significant about the strong correlation of temperature and atmospheric greenhouse gas concentrations is not which came first, the chicken or the egg, but rather that they are strongly linked. As with malaria and poverty, the relevance of this half-million year climatic march together, as recorded in the Antarctic ice, is the lesson that when one changes, so does the other. In the present day context, when the greenhouse gases in the atmosphere have increased by more than 30% over pre-industrial levels, the lesson we should learn from the past is that we can expect this enhanced greenhouse to be accompanied by increases in the temperature.

IDEOLOGICAL DISTORTION

Reconstruction of human history is burdened with the same difficulties as natural history: incomplete inaccurate, and conflicting information. But sometimes there is another factor that adds an additional distortion. On occasion, human history is degraded by purposeful suppression or deletion of available information. Such distortions arise when history is forced to pass through an ideological filter, to emerge in a sanitized form that conforms to a desired national image. In Japan, some school history textbooks still neglect to report the atrocities committed by the Imperial military during the occupations of Korea, China, the Philippines, and elsewhere in southeast Asia during World War II. The forcing of women to provide sexual services to the Japanese military forces, the use of germ warfare, and the brutal treatment of civilian populations in areas of occupation are facts of history that those who commission Japan's history textbooks would prefer to ignore.

Similarly, the sordid history of the Soviet Union under the dictatorship of Josef Stalin was for many years sanitized by government authorities anxious to overlook the shortcomings of the Bolshevik regime. Only after the dissolution of a monolithic one-party government in 1989 did Russians begin to learn in school about the brutality that characterized their government in the three decades following the 1917 revolution. And for most of the twentieth century, history textbooks in America largely overlooked the unsavory aspects of western expansion, the annexation of native lands, and the forced relocation of American Indian populations. Indeed, history can suffer distortion not only from an incomplete record but also when historians apply a selective filter. Science too can suffer when forced through an ideological filter. If biology teachers must conform to creationist ideology, as in the famous case of John Scopes in Tennessee, an understanding of biological evolution will surely suffer.

The past always merges with the future. Today is the tomorrow you worried about yesterday.[7] For processes that continue across the boundary of the present, an understanding of the past is a key to projecting into the future. But the past is a reliable key to the future only when processes or circumstances are unchanging with time. Geology and history, however, tell us that a static world is illusory. The future is related to the past, it is built on the past, but it seldom is an exact re-run of the past. In the next chapter, we explore the uncertainties and perils of predicting the future.

[7] This clever phrasing is attributed to Jerry Longan.

10 Predicting the future

> It is a mistake to try to look too far ahead. The chain of destiny can only be grasped one link at a time.
>
> Winston Churchill

Predicting the future... how enticing a prospect. Predicting the future has become a business for many. We can find, with only a little effort, fortune-tellers, clairvoyants, palm readers, astrologers, mystics, seers, psychics, and many others who will gladly reveal the future, for a price. But we all should be more than a little skeptical that any of these occult practitioners have special access to the future. Even economists and professionals who use less mystical tools – climatologists, actuaries and pension fund managers – find the distant future obscure and the pathway to it full of potholes. A principal theme of this chapter is that the future is a moving target, that divining its characteristics is always tough, and that it gets tougher the further ahead one tries to see. My philosophy for dealing with such uncertainty is to develop a long-term vision and make plans to move ahead – but to be prepared for many course corrections along the way, as the future unfolds quite differently than you have anticipated.

We all have heard that the only things certain about the future are death and taxes. This favorite adage surely captures the truism that most of the future is filled with uncertainty. The uncertainty is not uniform, however, and some aspects of the future are clearer than others. In addition to death and taxes, we can be reasonably certain that we will not win the lottery and that the sun will rise tomorrow.[1] Most other aspects of the future lie in the grayer precincts of the uncertainty spectrum.

[1] This regularity in the Earth's rotation about its axis was apparent even to Shakespeare when he wrote "...and it must follow, as the night the day..." (*Hamlet*, Act I, Scene III).

The truth of the matter is that it is really very difficult to predict the future, particularly very far into the future. Many predictions made not long ago are today recognized to be so far off target as to be ludicrous. Consider the following remarks from respected professionals of their time:[2]

Airplanes are interesting toys, but of no military value.
Ferdinand Foch, Commandant and Professor of Strategy, French War College, 1907–1911

Stocks have reached what looks like a permanently high plateau.
Irving Fisher, Professor of Economics, Yale University, 1929

I think there is a world market for maybe five computers.
Thomas Watson, Chairman of IBM, 1943

We don't like their sound, and guitar music is on the way out.
Decca Recording Company, rejecting the Beatles, 1962

We may smile at these examples, but they are more representative of most long-term projections of the future than we may care to think. The field of the future is strewn with casualties such as these that failed to see the future with greater clarity. Some, however, made mid-course corrections that let them ride the wave of the future. The history of the twentieth century reveals that IBM discovered there was a larger market for mainframe computers than Thomas Watson had anticipated, and that there was a grand future for small individual computers that anyone could learn to use effectively in their home. IBM's introduction of the personal computer in the 1980s was a course correction of enormous proportion.

SEEING THE FUTURE

One can find advice on just about any aspect of the future: which horse will win at the Kentucky Derby or Royal Ascot, which Nasdaq

---

[2] These four quotations come from a longer list displayed in the Smithsonian Institution's five-year (2001–2005) traveling exhibit "Yesterday's Tomorrows: Past Visions of the American Future".

or FTSE stock is about to leap ahead of the pack. The advice comes from those who study the reasons that horses or stocks have moved ahead to become winners in the past. Some horses perform better in shorter races, some on wetter tracks, some in less crowded competition, some on cool days, some with certain jockeys. Stock analysts likewise have developed ways of evaluating enterprises: the quality of and demand for a company's product or service, the strength of the management, the debt burden it carries, the loyalty of employees, the strength of the competition. When we buy tips on either horses or stocks, at some level we understand that there is risk and uncertainty in such decisions, and usually when the predictions do not pan out, we accept the consequences of taking risks and perhaps look for other counsel on such matters. The fact that different people will select different horses to win a race is at the heart of the old saying "That's what makes horse-racing". Every individual bet placed at the racetrack, or every individual purchase or sale of a given stock, reflects an evaluation of the likelihood of some future event. That we have a vibrant horse racing industry and vibrant stock markets attests to the willingness of individuals to face an uncertain future, exercise their best judgment, and take actions in the face of that uncertainty.

Another familiar look at the near-term future is the weather forecast. In this activity, the professionals at work have studied atmospheric physics and fluid dynamics, and they employ large fast computers to assimilate hourly information streams from meteorological observatories and orbiting satellites. Projections such as the overnight weather forecast, even the five-day forecast, have matured to a stage of reliability whereby they play an important role in travel planning, agricultural management, forest-fire containment, and warfare. But a five-day forecast of global financial markets with a similar high probability of being on the mark is not even a remote possibility on the horizon. This difference between meteorological and economic forecasts is simply a reflection of the fact that we understand meteorology far better than we do economics.

Each of these examples, horse racing, weather forecasting, and investing, involves a different time scale over which the uncertainty extends. The winner of the horse race is known within a few minutes after the race begins, and an evaluation of one's prior information and advice can take place. And meteorologists update their computational models with new data hour by hour and make revisions to their forecast daily. The success of an investment may not be apparent for days, months, or even years. But as the future unfolds, investors must be prepared to make adjustments to their portfolios to get back on a promising track, if the earlier pathway has led into a minefield.

Longer-term projections of what the local meteorological conditions will likely be a year or two hence are effectively out of reach, because many regional factors that affect conditions from year to year, such as the El Niño Southern Oscillation in the equatorial Pacific or the North Atlantic Oscillation in the northern high latitudes of the Atlantic, are themselves variable in duration and difficult to predict. Moreover, each of these regional features are coupled to the global system through teleconnections; a strong or weak El Niño in the Pacific Ocean is felt a year later in the precipitation over Zimbabwe. An estimate of global *average* conditions a decade hence, however, is less arduous, because one is asking far less of the projection into the future; in a global average projection the regional and short-term temporal details 'average out'. In a sense, one is seeking only the broad outline of the future, not the region-by-region and day-to-day detail. One is asking questions about the climate, not the weather.

Longer-term projections of course are made in economics as well. We hear confident predictions that a recession will last only ten months before economic growth is restored, or that interest rates will trend lower over the coming year because of changing situations in countries with developing economies. As the months and years unfold, the success or failure of such predictions often seem indistinguishable from a random walk. By contrast, a long-term climate forecast, an estimate of the average conditions to be expected a decade hence, is probably on much firmer ground than an economic forecast over similar time scales.

The shortcomings of both short- and long-term economic projections fundamentally derive from incomplete knowledge and understanding of the complex processes and factors affecting national and global economies. Weather forecasting, by contrast, is easy. Atmospheric scientists actually understand rather well the physical processes governing the short-term behavior of the atmosphere, so that with an array of meteorological stations and satellites feeding observations of solar radiation and atmospheric temperatures and pressures into large and fast computers, we get a reasonably accurate weather forecast. But what can we say about the processes governing the behavior of the stock market? What economic or behavioral laws govern the daily fluctuations and longer-term trends in this complex financial arena? These are much more obscure than the laws of physics, and accordingly there is a wide range of opinion and a great deal of uncertainty as to what will happen tomorrow, next week, or next year in the realm of national and global economies.

Each market analyst identifies what he or she believes to be the governing criteria and makes individual calculations about the future based on these criteria. Each has, in effect, constructed a personal model of market behavior to predict the future. Each prediction is what we call 'model-dependent', i.e. the outcome is dependent upon what the person making the prediction understands and believes about the processes and factors affecting market evolution. To the extent that the analyst has correctly identified relevant and significant criteria, the projection of the future will be more or less accurate. This uncertain 'wisdom' is for sale, and investors must exercise their own judgment or intuition as to which opinions have merit. Some might even calculate from many opinions a consensus, or 'average' projection for the future, as if each opinion were an inaccurate measurement or estimate of the economy's future status.

The great uncertainty about future market levels stems fundamentally from an uncertainty about the processes and factors that control the market. This is the uncertainty that I earlier referred to as the uncertainty arising from the *conceptualization* of the problem at hand. In the case of market behavior, the uncertainty of conceptualization far exceeds any uncertainties in the measurement of economic factors such as employment levels, inventories, factory output, and the like.

GETTING IT WRONG

With benefit of hindsight, we know that projecting the future is frequently off the mark, and unanticipated minefields are not uncommon. One of the principal newspapers in the area that I live not long ago ran the headline "Growth flooded freeways sooner than projected".[3] The accompanying story noted that traffic volume projections on a recently completed link of the interstate highway system north of Detroit were so far off target that the current traffic volume was already 50% greater than what had been projected for even a decade in the future. The unanticipated volume was attributed to unexpected economic and residential growth. In discussing how traffic volumes are estimated, the director of transportation programs for the regional council of governments said, "The assumptions going in don't necessarily translate into reality. The target never stops moving. One boardroom decision to move [the location of] a company can change things."

Clearly, estimating traffic patterns and volumes decades into the future is enveloped in uncertainty. Will the planners trying to cope with the new congestion learn from the previous experience, or will they simply repeat the mistakes of the past? The particular solutions discussed in the article included adding more on–off ramps, and more lanes of concrete. But at some future level of congestion, the public will realize that the basic conceptualization of the problem is probably wrong. The problem should perhaps be viewed as how to move people from where they are to where they want to go, rather than how to move automobiles. Once the planners and decision-makers have been freed

[3] *The Detroit News*, 2 July 2000.

from thinking in terms of automobiles as the only mode of transport, other possibilities can be considered. In Washington DC, congestion reached a stage thirty years ago where a subway transport system became an attractive alternative. Today, the Metro in Washington is the preferred mode of transport into, within, and out of the city for hundreds of thousands of people each day. Most large cities of the world – London, Paris, Moscow – long ago recognized the need of a metro system for people to reach the city center for their jobs or shopping.

Another story about walking headlong into a minefield has as a central character the mathematical model of the securities and financial markets developed by the Wall Street firm called Long-Term Capital Management (LTCM). This model served as the philosophical and technical underpinning of LTCM's investment strategy from 1994 to 1998. LTCM had developed an image of infallibility in part because its founders included two recent Nobel Prize winners in economics, and it had produced remarkable returns on investment in its first few years. It attracted very large sums of money from wealthy investors. Its fundamental investment strategy was to identify through intricate proprietary calculations a very small price differential between investment products that in an equilibrium market should have been priced the same. LTCM would execute massive transactions that would take advantage of these temporary small departures from an equilibrium market.

The strategy, while sounding attractive, faltered when a fundamental model assumption failed to materialize in reality. What was the flaw? LTCM's computer models assumed that there would always be a market: when you wanted to sell, somebody else would be there to buy. But in October of 1998, in the panic following the mid-August default on internationally held bonds by Russia, there suddenly were few buyers that would enable LTCM to complete the paired transactions. LTCM had bought but discovered that the opportunities to sell had vanished. Moreover, much of what LTCM had to sell was purchased with borrowed money, and the lending banks demanded payment. To repay the loans, LTCM was forced to liquidate other investments at discounted prices. Billions of dollars were lost by investors, and the looming inability of LTCM to repay its loans brought a rescue effort by the lending banks, other financial institutions and the Federal Government. The rescue team worried that a failure by LTCM might quickly lead to bank failures, and the subsequent financial ripples might turn into an economic tsunami with consequences beyond imagination. The lessons learned were many, and certainly the private investors who lost so much have their own long lists. But one important lesson for the wider public was that the banks should have been much more attentive to the fragility of their loans. That a private investment firm could undermine the stability of major banks should have been a wake-up call to loan officers and those who write the rules governing highly leveraged investments.

Another minefield was being laid out, unintentionally of course, at the very time the LTCM debacle was underway. In January 1998, following extensive deliberations, the state of California restructured and deregulated its electric power industry. The new structure separated power generation from power delivery, in effect breaking up the historic structure of the state-regulated vertically integrated electrical utility industry. Just before the restructuring took effect, a framer of the new legislation optimistically predicted "There will be an entirely new regime in which power is traded in an open market.... Five years from now, the retail price of power is going to be about half of what it is now." Three years later, he was on track for being both right and wrong. Yes, there was an entirely new regime in which power was traded in an open market, but it was a chaotic market. In those areas where deregulation had occurred,[4] the retail price, far from being reduced by half, had increased. But the wholesale price had skyrocketed to astronomical levels, and profiteering by energy generators was allegedly rampant. The California governor and legislature shaped emergency stopgap legislation to prevent the bankruptcy of the

---

[4]Not all of California underwent deregulation. Some municipalities, most notably Los Angeles, retained municipally owned and regulated electrical power generation. These areas were shielded from the gyrations experienced in deregulated areas of the state.

transmission and delivery segment of the restructured industry, which was being badly squeezed by the energy generators on the one side and unhappy customers on the other. But before the year of shortages ended, a major energy trading company and California's largest utility were both bankrupt. The dust has not yet settled on this episode, as new legislation is crafted to correct the flaws in the initial plan. A mid-course correction was definitely required.

As other states contemplate or start to shape utility restructuring plans, the California episode should yield lessons to be learned. Utility deregulation is a process in its infancy, and it should not surprise us that mistakes will be made. But infants learn from their mistakes, and so might we hope that consumers, providers, and legislators will likewise benefit from the California debacle. We may not be able to predict the future with any helpful degree of certainty, but we should be able to recognize when the future is unfolding in a deleterious fashion, and make the necessary course corrections to steer the electrical power industry into less turbulent waters.

DEFICIT OR SURPLUS?

Not all futures lead through minefields. For as many years as I care to remember, the tax revenues used to fund federal expenditures in the USA seemed inadequate for the projects and programs that Congress and the President deemed necessary. As a result, budget deficits became a perennial feature to be financed by governmental borrowing, and the national debt changed in only one direction: upward. As recently as the early 1990s, all of the nation's best economists were predicting ever-growing budget deficits, and all of the nation's best politicians were struggling to create balanced budgets that would, if achieved and maintained, stabilize but not reduce the national debt at some titanic level.

Early in the first year of the twenty-first century, all these gloomy projections had been set aside. The remarkable success of the US economy in the last few years of the twentieth century brought tax revenues to unprecedented levels, and the national treasury filled to overflowing. The talk turned to what to do with a huge *surplus*, and how fast the nation might repay its national debt. In late June of 2000, the revenue surplus (both social security and non-social security components included) was estimated to be almost $4.2 trillion in the first decade of the twenty-first century,[5] an amount greater than the entire national debt of $3.5 trillion. But what is particularly telling, at least in the sense of how much uncertainty there is in the economic forecasting business, is that this estimate of the decadal surplus represented an increase of more than 30% over an estimate made only five months earlier. And by early 2001, only six months later, the estimate of the surplus had grown again, to $5.6 trillion,[6] an increase of more than 70% over the projection made eleven months earlier. Halfway through the year 2001, however, the euphoria of revenue surpluses had already been blunted by a downturn in the global economy and the effects of a substantial politically motivated retroactive tax cut. And then came 11 September 2001, the terrorist attacks on the World Trade Center in New York. The response to that tragic event included massive federal funding for the reconstruction of New York, the subsidization of the air transport industry, the mounting of additional security measures at airports, and a war in Afghanistan. By year's end, the federal budget projected a deficit for the next three years.

Quite apart from the surprise of the terrorist attacks, it is abundantly clear that it very difficult to model the national economy, and therefore there is great uncertainty in the projections of federal revenues. But this uncertainty does not allow the luxury of simply waiting for the situation to clarify. Budgets must be formulated, spending decisions must be made, revisions to the tax code considered, all in the face of the uncertainty. No matter whether we embark on a path of debt reduction, tax reduction, or a re-prioritization of programmatic spending, or experience new surprise attacks, we must recognize that in a few years hence the nation may need a course correction as conditions change. This concept was introduced into the Congressional

[5] *New York Times*, 27 June 2000.   [6] *New York Times*, 31 January 2001.

debate about cutting taxes, where it was termed a trigger clause or safety valve that would interrupt tax cuts if the revenue did not materialize as anticipated. It was not adopted.

RUNNING OUT OF OIL?

Stephen Jay Gould observed[7] that "Almost all our agonized questions about the future focus upon the wriggles of the short term, rather than on the broader patterns of much longer scales." But how should we react to, how can we address the uncertainties that extend perhaps beyond our lifetimes? One such question that arises from time to time is, "When will the world run out of oil?" The wide array of answers to this question is bewilderingly diverse. On the one hand, we hear optimists say, "Never"; on the other, we hear that this century will certainly be the last in which oil powers the global economy. One thing about oil is certain, however. The rate of creation of oil in the Earth proceeds on the slow geological time scale, whereas its consumption is taking place on the fast human time scale. In other words, nature will not rescue us by making oil as fast as we are extracting it. Whatever nature has created over the long history of the Earth is the resource that we have to work with.

Many factors, of course, will affect how long petroleum products will continue to fuel motor vehicles and electric power stations, heat homes and office buildings, and provide feed stocks to the petrochemical industries. The demand for energy derived from oil may be tempered by energy conservation measures such as better insulation of structures, higher fuel efficiency in household appliances and vehicle engines, smaller lighter cars, better public transportation options, and of course the development of alternative fuels. Coal and natural gas are widely available globally, and nuclear energy already is used to generate some 10 to 15% of the world's electricity. In some areas, solar heating, solar electric power, wind, hydroelectric, and geothermal energy are also making substantial contributions. To be sure there are

---

[7]This View of Life, in *Natural History*, September 1998.

environmental issues associated with many of the alternatives to oil (and of course with oil itself), but in the simple calculus of adding up kilowatt-hours, there are ample sources of energy that can and will reduce the demand for oil.

Cost will also affect the demand for oil. Oil has been the fuel of choice throughout the twentieth century because it has been cheap. In real (uninflated) prices, the cost of petroleum-derived fuels actually declined throughout the twentieth century. When I first began driving in the 1950s, Dwight Eisenhower had just been elected president of the USA and the price of gasoline was twenty cents a gallon. A dollar in the tank was good for a busy Saturday night on the town. Today the price sits somewhere between one and two dollars per gallon, about a seven- or eight-fold increase in the price over the intervening half-century. What else can you remember that has not increased far more? Housing, whether purchased or rental, is up by more than a factor of ten. As a university student, I could budget one dollar per day for food (admittedly a rather spartan diet), whereas today ten dollars per day can do little better. Oil, however, has remained relatively inexpensive. For many decades, oil traded for three dollars a barrel, and except for short intervals of time driven by international conflict, it has remained well below thirty dollars per barrel to the present day.[8]

The price of oil has risen more slowly than the general cost of living because it has been abundant. The supply of oil has been able to grow as the demand has grown, thus tempering the price of gasoline at the pump. The big international oil companies remind us that they have not let us down in the past and are fully capable of sustaining us into the future. But will the supply of oil continue to be adequate to meet the ever-rising energy needs of the modern industrial economy? The answer to this query is at the heart of the question of when the world will run out of oil.

Part of the diversity in the answers to this question stems from the definition of oil. The most common image of oil is that of a liquid

[8]Only the minimum wage, that entry-level salary for workers with no special skills, has increased less: from 75 cents in 1950 to $3.75 in 2001, a five-fold increase.

residing in the rocks below the Earth's surface that can be induced to flow to a well from which it can be pumped to the surface. However, in addition to the liquid oil there is also oil in solid form that will not flow and cannot be pumped. The solid oil, in the form of oil shale or in gooey bitumen locked in the pores in sandstone, greatly exceeds the liquid fraction but is not easily recovered. The assertion that we are likely never to run out of oil usually derives from a consideration of the total oil resource, both liquid and solid. What is usually left unsaid is that the cost of extracting the solid fraction will be much greater than the relatively easily pumped liquid fraction. In fact, the technologies for extracting the solid oil are in their infancies and are themselves much more energy intensive than the relatively simple drilling into and pumping out the liquid oil. So while there is a large amount of oil still resident in the Earth, most of it is not easily within our grasp nor will it be soon.

What can we anticipate in terms of the liquid supply? Contrary to popular belief, and contrary to what many oil producers tell us, we have already seen the finiteness of this resource. The fundamental reason behind this assertion is that we are extracting more oil than we are discovering. That is an equation that leads inevitably to exhaustion of a resource. When more fish are taken from the sea than can be replaced through natural reproduction, the fisheries decline as a resource. When more water is pumped from subsurface aquifers for irrigation than is replenished by infiltration from the surface, the aquifers decline as a resource. Oil is no different. The world is currently consuming some 26 billion barrels of oil each year, and discovering only about 6 billion barrels of new oil each year.[9] When we consume more than we can find, we are on a pathway to the end of the resource. And the statistics behind this shortfall are very clear and unambiguous.

In response to increasing demand, production of oil in the USA increased decade by decade through the twentieth century until it peaked around 1970. It has declined since then, and today stands at about 80% of the 1970 peak. That fraction will continue to fall, no

---

[9]Walter Youngquist, *Geotimes*, pp. 24–27, July 1998.

matter the intensity of exploration and technical innovations in exploitation. The simple fact is that oil is becoming harder to find and harder to get out of the ground. The easy days of discovery and production appear to be behind us.[10]

That is not, however, the story that you will hear from the oil producers and their friends in Congress. They tell us that if they are freed from environmental constraints as to where they can explore and drill, there is ample oil to be found and produced.[11] But the decline in production in the USA over the past three decades has not come from arbitrary restrictions, nor has it come from a paucity of technical innovation. To the contrary, the decline has taken place even as the geography of exploration has expanded to the continental shelves and deep ocean, and the art of exploration has advanced dramatically through the use of supercomputers, three-dimensional imaging and visualization of what lies beneath the surface. Production of discovered oil has declined even as new drilling and extraction technologies have improved substantially. Without discovering oil in new geological environments and without advances in production technology, the decline from the 1970 peak would have been even swifter.

Demand, of course, has not tapered off at all. At the peak of domestic production in 1970, Americans were consuming about fifteen million barrels of oil each day, and the domestic production provided more than three-fourths of that consumption. But today, the consumption has grown to more than twenty million barrels of oil per day, while the domestic production has continued to fall. That has led to an increasing reliance on imported oil to meet the shortfall, to the point where in 2001 the USA imported 56% of its daily consumption, compared with only 23% in 1970. This dependence on foreign oil will only grow as time passes. In this context, the energy policy of

[10]The history of oil production and its implications for the future is the topic of the book by Kenneth Deffeyes, *Hubbert's Peak: The Impending World Oil Shortage*, Princeton University Press, Princeton, NJ, 2001, 285 pp.

[11]In response to a question from a reporter about the supply of oil and gas in the USA, a leading figure in the US House of Representatives replied "We have an unlimited supply. We just haven't found it".

the second Bush administration had as a target the reduction of imported oil to only 50% of national consumption, principally through increased domestic production. To meet that goal would require an increase of domestic production to a level not seen since the early 1980s, in effect a reversal of the long downward trend in production that would defy the historical experience. It is a prime example of wishful thinking.

What has already happened in the USA will soon overtake the global production of oil: we will see worldwide production peak early in the second decade of this century. When that occurs, sometime between 2010 and 2020, the world will be embarking on a new economic pathway. At that point, we will see declining production failing to keep pace with increasing global demand. Approximately half of the liquid oil resource will still be in the ground, but harder and harder to extract. The outcome will be clear enough. Prices will increase substantially, and demand for oil (but not for energy) may be curtailed as the reality sinks in that the days of cheap abundant oil are behind us. The solid oil will beckon, but the technologies to extract this resource will not likely be in place to ensure a smooth transition away from conventional liquid oil. Initially reliance will probably shift to natural gas, a fuel with a production and depletion history similar to oil, but delayed in time. Alternative energy scenarios could also, in principle, ensure a smooth transition away from oil, but the pace of development of the alternatives must accelerate if they are to take up the slack when conventional oil production begins to decline globally in a decade or so.

CLIMATE OF THE FUTURE

Another uncertain future, for which outlines are only now beginning to emerge from the haze, relates to the warming of Earth's surface from an array of natural and anthropogenic (i.e. human-induced) causes. The uncertainty has focused on many different aspects of this topic at different times in the evolving debate (in the last chapter I will take the opportunity to review climate change in detail). Initially, the data offered to support the assertion that Earth has been warming was challenged as being insufficient and inaccurate. Later the debate turned to whether the warming, by then acknowledged as real, was significantly outside the range of the natural ups and downs of the climate as deciphered from the historical, archeological, and geological records. When additional research indicated that the rise in temperature over the twentieth century was indeed unusually large and fast, the argument shifted to the causes of the warming.

The overwhelming scientific consensus presented by the IPCC (Intergovernmental Panel on Climate Change, an organization first mentioned in Chapter 5, working under the umbrella of the United Nations and the World Meteorological Organization) acknowledged significant roles for natural factors such as variable radiation from the sun and occasional perturbations from volcanic eruptions, and for human factors such as the increase of greenhouse gases, and particulate matter and aerosols in the atmosphere from the burning of fossil fuels. The scientific studies indicate that over the past millennium the natural factors dominated climatic fluctuations up to about 1750 or 1800. From that time until about 1950, the human factors grew to a sufficient potency to rival the natural factors, leading to a climate variability derived from a very complex blend of forcings. In the latter half of the twentieth century, however, the human forcings outpaced the natural factors by a large margin, and the human fingerprint on the warming of the planet has become ever more apparent. But the cloud of uncertainty, some truly scientific and some deriving from the smoke-screens floated by industrial self-interest, has confused the public and made people cautious in accepting the scientific underpinnings of global climate change, both in the past and as projected for the future.

Forecasting Earth's climate over the next century is not an easy undertaking. The natural and social scientists who study this broad issue acknowledge that many complex and interdependent factors will influence evolution of the climate. For starters, how many people will be added to Earth's population, and where will they be living? What will their standard of living aspirations be? What new energy sources will power the global economy, and when will they be phased in? Those who pretend to know the answers to all of these questions are either deceiving themselves or trying to deceive you. In truth, each of these questions is a highly complex issue, with many possible answers. In fact there is far more uncertainty about these factors than there is in the understanding of the way the climate system of Earth works. The meteorology and climatology, as expressed in the many equations of physics and fluid dynamics to be solved on large computers, is on far firmer ground than the social and economic trends that give climate models the data they need to calculate the climate a century or more into the future.

The scientists of the IPCC have provided a range of scenarios about the future that might be the outcome of certain demographic and economic pathways through the twenty-first century. One family of scenarios describes the century as a period of very fast economic growth, the rapid introduction of new and more efficient technologies, and a population that peaks in mid-century and declines thereafter. This family of scenarios is characterized by an economic convergence between regions and increased cultural interactions, a true 'globalization' of the economy and a blurring of cultural and regional distinctions. Within this general framework are multiple pathways that depend on the energy choices that will drive the global economy: fossil fuel intensive, non-fossil intensive, or a diversified blend of all energy sources available.

A contrasting family of scenarios envisions a much slower trend toward globalization, and the maintenance of a more heterogeneous world. Fertility patterns remain distinct and disparate, with some regions showing large population growth while others stabilize. Economic development is regionally focused, technological changes diffuse more slowly, and energy usage remains largely dependent on fossil sources.

From each pattern of economic development, population trajectory, and energy choices comes a projection, via complex model calculations, of greenhouse gas concentrations in the atmosphere and the accompanying increase of the global mean temperature. These should not be thought of as individual forecasts of climate evolution over the twenty-first century but rather as an ensemble of scenarios that define the likely range of climatic conditions we will encounter as the century unfolds. It is probably far more useful to know the likely range of possible outcomes under widely different assumptions than to quibble about the details of any single scenario. The procedure can be envisioned as a systematic and methodical exploration of outcomes, a series of numerical experiments with different values of the experimental variables. The IPCC climate forecast for the twenty-first century should be seen both as a process and a product.

The strategy of comparing and evaluating multiple scenarios is not unique to forecasting the climate of the future. It is a very common approach in assessing the condition of the economy of the future. One can read reports like, "The economic models of sixteen investment banks indicate a range of growth between 2.3 and 4.1% over the next year", or, "twenty-four analysts specializing in the bond market predict that interest rates will diminish by 1 to 3% over the coming year". As with climate scenarios, it is often more useful to have a feeling for a *range* of economic outcomes defined by many independent assessments than to give undue emphasis to the intricacies of a particular model.

Such an examination of the future by the systematic evaluation of multiple scenarios is a variant of T. C. Chamberlin's Method of Multiple Working Hypotheses discussed in Chapter 9. Decisions about the future are difficult only because of the uncertainty that envelops them, and because the uncertainty will not likely be resolved in a helpful time frame. Thus decision-making necessitates an evaluation of many options, an assessment of the consequences of different options for the future. The *status quo* is usually an option, one that is governed by inertia. That option is usually favored by those with a vested interest in preserving the *status quo*, and who will be adversely affected by decisions that deviate from it. In evaluating many options, including both the *status quo* and those involving dramatic change, we must understand the philosophical or economic ground from which the various options stem. Who will be the winners and losers in each scenario? What weight shall we give to arguments stemming from obvious self-interest? Will a reasonable delay in making a decision allow new information to emerge that will reduce uncertainty? Can we buy 'insurance' that will buffer adverse outcomes in an uncertain future? All of these questions can be usefully addressed and can illuminate pathways through the thickets in the garden of uncertainty. If the garden occasionally captures us in a maze, we may need to back out of some blind alleys and dead-end corridors. As I mentioned at the beginning of this chapter, it is difficult to see the path to the future clearly. It is easy to be wrong about the future, but that should not deter us from starting to make our way through the maze, recognizing, of course, that some mid-course corrections will likely be necessary.

As a strategy, 'multiple working hypotheses' forces one to think about a range of futures, a range of possible pathways. It forces one to keep an open mind, to evaluate options and to consider course corrections. Remaining open to ideas and fresh ways of thinking about a problem is the very antithesis of succumbing to an ideological perspective. Ideology has appeared frequently in the garden of uncertainty: in Chapter 2 it emerged in the context of the sowers of uncertainty; in Chapter 7 it was seen in the context of conceptualization and 'thinking outside the box'; and again in Chapter 9 it appeared in the context of how ideological filters distort history. Falling under the spell of ideology is equivalent to narrowing the family of possible scenarios to a single pathway. A closed mind will see only one pathway, and it will likely turn into a deep rut eventually.

In this chapter I have tried to make the case that the future under many circumstances cannot be seen very clearly, and it becomes even murkier the further one tries to see. Course corrections will be necessary to escape from ruts before they are too confining. But sometimes our vision of the future is impeded not just by dim illumination. On occasion, the window to the future is totally opaque. These are the times when we are blind-sided, when an event occurs that takes us totally by surprise. Such an event occurred on 11 September 2001. Such events are the topic of the next chapter.

11 Out of the blue

Expect the best, plan for the worst, and prepare to be surprised.
Denis Wheatley

An event that cannot be anticipated, an occurrence we might call a random happening, is inevitably a source of uncertainty, if for no other reason than we do not know when it will occur. At a personal level, we may be familiar with many such events: an automobile accident, a fire or burglary at home, the sudden death of the family breadwinner, and sadly, even a surprise terrorist attack. But there are other misfortunes that can occur on a regional or even global basis, equally sudden, equally catastrophic. We are familiar with such events also, although perhaps not through direct personal experience: a flash flood, a widespread power failure, a devastating earthquake, a major volcanic eruption. Whether personal or more widespread, all are events that we fervently hope never to experience. Yet we know that on any given day there is a small, but not infinitesimal, probability that our lives may be touched by one or another of such events. Accordingly, we have developed some common strategies to deal with their consequences. The mechanisms we employ to cushion us from the consequences of these unwanted and unanticipated events include *emergency preparedness* and *insurance*, both examples of our tendency to look to the future with caution.

EMERGENCY AND DISASTER PREPAREDNESS
In the immediate and personal environment of our home, we often take steps to forestall catastrophe. We install smoke detectors to provide early warning of fire, and perhaps to trigger a sprinkler system to quench incipient flames. We may have an in-house security alarm system that will automatically call the fire department or police if the system is tripped by signals of fire or unauthorized intrusion. We

may keep a supply of batteries, bottled water, and comestibles to tide us over in the event of a prolonged power outage. We may even have a small stand-alone electrical generator that can be fired up to provide emergency power. Or we may choose to do none of these, and just take our chances with whatever the future may bring.

Similarly, there are actions that can be taken collectively through the decisions of governmental bodies. Zoning laws can restrict residential construction in flood plains, and building codes may require higher standards of construction materials and design in areas prone to earthquakes. Municipalities may designate and supply emergency shelters in the event of earthquake, flood, or hurricane dislocation. The nation may establish a national petroleum reserve as a fuel supplement in times of emergency when external sources may be interrupted.

Unanticipated events that might impair very complex and highly distributed systems such as the electrical power distribution grid require special considerations. Redundancy in function and design is one common approach. Redundancy in design is now central to our thinking about how to prevent total failure in complex systems. In some ways, redundancy is the opposite of the 'domino effect', in which one domino in a long row falls, toppling the next, and the next. The chain reaction ends only when all the dominoes have fallen. In a system with sufficient redundancy, a fault in the system at one place can be isolated and bypassed, with other parts of the system taking on the role of the defective element, thereby preventing a system shutdown.

We all have at one time or another in our travels come upon a road or bridge closure and have been forced to follow a detour leading us around the problem, eventually taking us to our destination. A well-designed transportation network usually has sufficient redundancy to avoid the creation of bottlenecks, let alone a total shutdown. A well-designed electrical power distribution system has sufficient redundancy to accommodate the occasional failure of a transformer, the inadvertent cutting of an underground cable, or the emergency shutdown of a generating plant. But occasionally all the dominoes fall, and inadequacies in the system are highlighted in the cascade of failures that result in a widespread blackout.

Today cities usually require utility systems – gas, water, electricity – to have many degrees of redundancy in their design, so that detours around damaged regional segments can be quickly activated. In fact, much of the temporary rerouting in electrical power distribution systems is relatively free of human intervention, with load sensors and self-actuating switching devices kicking into action when a fault disrupts the distribution network. Such anticipatory design seems rather obvious and wise, but it was not always so. Following the 1906 earthquake in San Francisco, in addition to the direct structural damage caused by the shaking, fire was a significant cause of subsidiary damage. The destruction was made worse because the city water supply system had been compromised by the quake. This led to insufficient water with which to fight the fires, because there was no way to cut off the supply to regions where ruptured water mains lost water that could have helped to extinguish conflagrations elsewhere. One can easily imagine how such a problem today might be further compounded by an inability to cut off natural gas to areas where gas pipes have been ruptured and are providing a ready supply of fuel for fire.

The value of redundancy in natural systems, particularly biodiversity, is also coming to be more fully appreciated. The intertwined character of an ecosystem sometimes becomes apparent only when we begin to disrupt it. In a natural and diverse system, there may be many pathways for carbon and water to move through the system in their respective cycles. However, as land-use changes are imposed on a region, the hydrological patterns are interrupted, and habitat is altered or destroyed, fewer choices for 'detours' remain. One of the first alterations that historically has accompanied land development is the draining of wetlands, pejoratively known as swamps. But the importance of wetlands as resting places in the annual migration of birds, and the value of natural wetlands as a purification and filtration system for natural waters, has only belatedly been recognized. As we inadvertently destroy links in the connectivity of a complex ecosystem, the system will experience a degradation of function, and it may ultimately collapse when no further detours can be achieved.

But how can scientists and engineers identify every weak link in a complex system, every potential bottleneck and domino that can wreak havoc in the real world in which we live? One approach is to undertake 'what if' numerical experiments to attempt to understand the consequences of such an event. 'What if' experiments fall in the category of large computer models (see Chapters 7 and 8) of very complex phenomena, in which there are many linkages and feedbacks between elements of the system. Random perturbations to one or another elements of the system are introduced, and the consequences observed as they propagate through the system. For example, electric utility companies play 'disaster games', simulating unplanned impairment of their generating capacity, or failure in branches of their transmission networks. By systematically introducing problems into a system at various locations or in various component processes, the experimenter or analyst can sometimes discover vulnerabilities. Computer simulations can examine very large numbers of possibilities and establish probabilities for various types of failure, each with different consequences.

One of the big success stories in trying to anticipate problems and make adjustments to obviate them was the saga of 'Y2K', the feared computer glitch that might have caused chaos in the world as the calendar rolled over from 1999 to the year 2000. In the early days of computers when data storage space was at a very great premium, many computer programs used a shorthand for entering the year, by truncating the first two digits. The year 1967 became 67; 1994 was represented as 94. But in the 1990s it became apparent that the turn of the century, the moment when 1999 turned to 2000, would introduce an ambiguity: would 02 mean 1902 or 2002? Systems and programs that depended on the calendar – systems involving scheduling, salaries, commodity futures, retirement pensions, the list could be virtually endless – needed to be fixed to avoid end-of-century turbulence. Because the exact moment at which the problem would manifest itself was always known, one measure of uncertainty was removed from the equation. Many person-years of effort and large sums of money were expended around the world to repair this shortcut taken in the early years of cyber history. That the calendar rollover occurred almost flawlessly was certainly the result of this massive effort at advanced remediation. Nevertheless, one also heard the opinion that the reason nothing significant happened at the rollover was that the problem and its likely consequences were portrayed to be far more consequential than they really were.

The response of political and social structures to major unanticipated events can also be explored through computer modeling, although quantitatively representing the relationships and interactions between various model elements remains a challenge. Nevertheless, strategic and military planners construct and play 'war games', testing the responses of various conceptualizations of social, political, and military structures to a variety of disruptions, which may include random events. In planning the Gulf War of 1991, many simulations were carried out on computers, testing responses to a variety of situations that might have developed. What if Iran entered the war on the side of Iraq? What if Libya launched a missile attack on Israel? How might a prolonged sandstorm affect the mobility of the US tanks and armored personnel carriers? What would be the consequences of a week of thick cloud cover on close air support of ground troops? What would be the effects on public support within the US for the war effort if casualties were mounting in a ground war of attrition?

Of course, one must first imagine possible scenarios before one can explore them in war games. The September 11 scenario of the deliberate crashing of fully loaded and fueled passenger aircraft into skyscrapers apparently had not been seriously considered or even imagined, although the experience with the Japanese kamikaze pilots in World War II certainly foreshadowed such a possibility.

Many computer games that young people spend countless hours playing hold their interest because they face different circumstances each time they play. Although the basic ground rules remain the same, the computer has altered the terrain each time. Similarly, card players can play endless hands of bridge, each time facing a new deal, new combinations of cards generated randomly by the computer. Similarly, computers now demonstrate formidable skill at playing chess, adjusting strategy following each move by a human (or computer) opponent. The better players are those who assimilate the lessons of these multiple experiences and who develop patterns of play that respond successfully to broad categories of circumstances.

'What if' experiments can also be conducted for natural systems. We might, for example, want to learn how Earth's climate system would respond to a major volcanic eruption. There is ample evidence in the historic and geologic record that a major eruption can have significant effects on climate. Following the eruption of Mt. Pinatubo in the Philippines in 1991, Earth's global average temperature fell by $0.3\,°C$ (about $0.6\,°F$) over the next two years as the ash and aerosols partially blocked the sunlight from reaching Earth's surface.

In 1815, Mt. Tambora violently erupted in Indonesia, in a massive explosion that was heard in Jakarta, some 1200 kilometers (750 miles) away. Tambora explosively injected more than 140 cubic kilometers (thirty-five cubic miles) of ash and pumice into the atmosphere, more than ten times as much as did Pinatubo 176 years later. The year following the eruption of Tambora, 1816, was known around the world as 'the year without a summer'. Temperatures were lower throughout the northern hemisphere, snow fell, lakes froze over at places in the New England states, crops failed widely in North America and Europe. The English poet Lord Byron, enduring a miserable summer holiday in Switzerland, wrote his poem *Darkness* in which he described the local effects of the Tambora eruption with these words:

The bright sun was extinguish'd, and the stars
Did wander darkling in the eternal space,
Rayless, and pathless, and the icy earth
Swung blind and blackening in the moonless air;
Morn came and went – and came, and brought no day...

In the sixth century, a major eruption also occurred,[1] which was widely noted in the historical records of Europe and Asia. The effects of this eruption, occurring in 536 AD, was described by a contemporary writer[2] in the following words:

The Sun became dark and its darkness lasted for eighteen months. Each day it shone for about four hours, and still this light was only a feeble shadow. Everyone declared that the Sun would never recover its full light. The fruits did not ripen and the wine tasted like sour grapes.

In a 'what if' experiment exploring the climatic effects of a volcanic eruption, the experimental variables would include the volume of volcanic ash and dust that is ejected into the atmosphere, the sulfur content of the gases and aerosols, and the explosiveness of the eruption, which would affect the altitude in the atmosphere to which the dust would reach. Other important factors are the location of the eruption and the season at the time and place of the eruption. An eruption in the temperate zone will have a significantly different effect than a polar or equatorial eruption because of the different patterns of wind circulation, and a wintertime eruption that spreads dark ash over an otherwise snow-covered white terrain will alter the balance of absorbed and reflected sunshine. A systematic exploration might experiment with eruptions in all of the well-known volcanic zones of the Earth, with ongoing steady eruptions versus single explosive

---

[1] Probably occurring on the island of New Britain, east of Papua New Guinea (Heming, R. F., *Bulletin of the Geological Society of America*, vol. 85, pp. 1253–1264, 1974).
[2] Probably John of Ephesus, but also attributed to Michael the Syrian (Rampino, M. R., Self, S., and Stothers, R. B., *Annual Reviews of Earth and Planetary Sciences*, vol. 16, pp. 73–99, 1988).

events, with the chemical nature of the gases emanating from the volcano, and with numerous other factors. The greater the number of scenarios examined, the more we will learn about the possible climatic consequences, and the more confident we will be about what to expect in different places and at different times.

INSURANCE

Insurance is a common concept, familiar to almost everyone who wants to protect something against possible disaster: automobile insurance, homeowner's insurance, health insurance, life insurance. We make an annual payment to an insurance company, which in turn promises to pay us an amount of money to cover our losses, or at least to enable a smooth transition to a different future, if disaster actually strikes.

Certain types of insurance are not even considered voluntary. Automobile liability insurance is one type of involuntary coverage. In some states, you may choose to have no protection for damage to your own car, but you must have some coverage to reimburse others who suffered damage or injury in the accident should you be at fault. In other states with so-called 'no fault' insurance, each vehicle owner is obliged to have insurance that comes into play no matter what the apportionment of fault may be. I call such insurance involuntary, because it is often a legal requirement, evidence of which must be provided, in order to obtain a license plate for one's vehicle. Another involuntary coverage is homeowner's insurance, at least in the case where a lending institution holds a mortgage for which the home is pledged as security. A usual condition of obtaining the mortgage is that the lender be protected from loss through insurance.

How is a premium set for a given type of insurance? How does an insurance company know how much to charge a customer who seeks coverage? The principal foundation of insurance pricing is an estimate of the likelihood of an insured event occurring. This likelihood is, in turn, primarily although not entirely based on past experience. Life insurance is paid upon the death of the insured, and, therefore, mortality statistics provide a baseline for the average life expectancy, as well as establishing the patterns of how many people suffer an early demise, and how many live beyond the average life expectancy. With sufficient numbers in the historical population, these statistics give a good estimate of the probability of dying at any given age, either earlier or later than the average.

Insurance companies and their actuaries (mathematicians with special skills in probability, statistics, and risk analysis as applied to the insurance business) are clever enough to recognize that there are variations in mortality in a general population that can be attributed to occupation. For example, coal miners and sky-diving instructors may pay more for life insurance than do school teachers or accountants. Smokers often pay more than non-smokers of the same age. Likewise, variations in life expectancy occur geographically, owing to regional environmental health factors, and ethnically, where variations in the predisposition to high blood pressure, for example, may lead to different mortality statistics among different ethnic groups. Moreover, the task is made more difficult because the actuaries must anticipate conditions many decades into the future. Will the past experience in mortality be a good estimate of the future? Surely improvements (or deterioration) in the economy, in diet, and in public health will influence future longevity.

Similar principles apply to the pricing of homeowners' insurance. Basic historical statistics inform today's price structure. How many homes are there in the community or county? How many home fire calls are received by the local fire department in a year? These simple quantities provide a baseline annual probability of home fire occurrence. Secondary considerations in pricing are also based on whether your property is more or less vulnerable than the average structure. What type of structure is most or least prone to catching fire? How far is your house from the nearest fire station or fire hydrant? And of course the premium will depend on the size and quality of your home and its furnishings, which are important factors in estimating its replacement value. Special considerations also apply to other hazards that homeowners may choose to insure against. Premiums may be higher in flood plains or in areas prone to earthquakes.

Just as life expectancy changes over time, so also do many factors that affect property insurance premiums. Insurance companies must be aware of and make estimates of trends over time. How will the rate of inflation change? What effect will a changing economy have on community property values? What effect will urban sprawl have on the pattern and frequency of flooding, when conversion of open agricultural land to paved streets and parking lots alters the natural pattern of water infiltration and runoff? Such assessments are often made by complex numerical models similar to those discussed in Chapters 7 and 8.

But just as weather forecasts are best for short intervals into the future, so also the demographic and economic factors needed to project insurance costs are best when estimated only a year or two beyond the present. Certainly, the insurance companies will be taking longer looks into the future, but the safest strategy for dealing with time-varying quantities is typically through an annual adjustment of both coverage and premium levels, a short-term periodic 'course correction'.

Beyond the common types of hazard insurance – life, health, property – that individuals and corporations buy as protection from an uncertain future, there are more specialized insurance products: agricultural crop failure insurance, ski resort snowfall insurance, business interruption insurance, earthquake and flood insurance, to name a few. Many can be associated with weather and climate phenomena.

Imagine a flood such as that which inundated much of the city of Grand Forks, North Dakota in the spring of 1997. Downtown merchants of course suffered property losses caused by flood damage, some of which was not insured. But their economic distress was amplified and prolonged by the fact that they were unable to conduct business or provide services for a substantial period following the recession of the floodwaters.

Or consider farmers who must plant crops in the spring and then watch and wait as the crops mature over the growing season. Without adequate and timely rainfall, the crop yield can be severely diminished or lost altogether. Similarly, proprietors of winter recreational areas depend on the availability and timeliness of snow. Snow-making machines can make a difference on the margin, but there is no substitute for abundant natural snowfall. Consequently, farmers and ski resort operators can protect themselves from the uncertainties and vagaries of the weather with specialized insurance.[3]

SURPRISES

The natural world surprises us with many unpredictable events – volcanic eruptions, earthquakes, flash floods. Every year, some 4,000 earthquakes large enough to do damage occur somewhere on Earth. And although we have learned a fair amount about the geography of earthquake occurrence, we still have not made any real progress in estimating *when* an earthquake might occur in a given region.

Earthquakes do not occur randomly over the face of the Earth, nor are they uniformly distributed over the globe. A century of observing *where* earthquakes take place indicates that they occur in very well-defined geographic zones. Americans living in California are all too familiar with the infamous San Andreas fault, and the citizens of Turkey have learned through misfortune about the Anatolian fault running across the north of that country. However, to pinpoint exactly when and where the next catastrophic event will occur along these fault lines is still beyond the reach of seismologists. Careful analysis of the timing of seismic events shows no regularity between occurrences; in fact, these analyses cannot reject the hypothesis that the time intervals between earthquakes, either locally or globally, is fully random. The next one will come when it comes.

To be sure, seismologists have learned something about recurrence intervals of major earthquakes. In southern California, along the San Andreas fault east of Los Angeles, geologists have determined that major earthquakes have disrupted the land surface in 1857

---

[3] For an interesting discussion of weather insurance, see the article by James Surowiecki in the *New Yorker*, 23 July 2001, p. 29.

(an event well-documented in historical records), 1745, 1470, 1245, 1190, 965, 860, 665, and 545. What we learn from this is that the time interval between these nine major earthquakes along this segment of the San Andreas fault averages about 170 years, but it has been as short as 55 and as long as 275 years. Not exactly a reliable guide to the future, although it gives some very tentative backing to the worry that, at least in southern California, a 'big one' may occur in the next several decades. But this very uncertainty about when and where the big one, or for that matter any damaging quake, will occur also allows people to build and live in homes in seismically active areas.

What protection is available for earthquake damage? With so much uncertainty about when an earthquake might occur, insurance premiums are set accordingly. In California, earthquake insurance is provided only by the California Earthquake Authority (CEA), a privately financed but publicly managed entity (with a motto that proclaims "We can't predict the future – we can only protect it"). CEA annual premiums vary with location but generally are in the range of 0.1 to 0.6% of the cost of the edifice (not the land), with a whopping big deductible of 15%. On a home valued at $300,000 the owner would pay the first $45,000 of repairs and reconstruction before the insurance would kick in. Clearly insurance here is protection only against total catastrophe, but it still leaves very substantial risk and responsibility to the homeowner.

Flood insurance is also available in California. At first glance, it is priced similarly to earthquake insurance, with annual premiums set at a small fraction of the home's value. For the same $300,000 house described above, both the earthquake insurance and the flood insurance will cost about $600 each year. But there is a *big* difference in the deductible. The flood insurance will require the homeowner to pick up only the first $500 of flood damage, whereas the earthquake insurance will leave a $45,000 deductible gap. For about the same premium, the homeowner is buying a whole lot more protection against floods than against earthquakes. The setting of such a high deductible for earthquake insurance is a recognition that geologists and insurers know much less about earthquakes than they do about floods; scientific uncertainty is reflected in the requirement that the homeowner share a much higher proportion of the earthquake risk.

Could the CEA offer a policy that had a smaller deductible? In principle yes, but of course at a higher price. Were the deductible to be reduced and the premium increased each by a factor of ten, one would be paying $6,000 annually for insurance with only a $4,500 deductible. Over 50 years, one would have paid the insurance company an amount equal to the value of the home, and at that premium level the policy might better be called a long-term self-financed reconstruction plan rather than insurance protection.

The insurance industry is in the forefront of risk assessment, trying to anticipate trends of the future. This industry cushions and buffers itself from the adverse effects of events both natural and anthropogenic, and at the same time it helps customers to face an uncertain future with contractual compensation for loss and damage. The insurance companies need to be among the first industries to recognize changing demographics, changing life-styles, changing construction standards, and changing economics. On the basis of these, they need to make periodic adjustments to the structure of their policies and the levels of their payouts and premiums. Because they are among the very first to pay the consequences of change, the insurance companies have less room to maneuver than some other industries, which can perhaps afford a more leisurely 'wait and see' approach to understanding the implications of change. Ideological denial of change is a perilous attitude for any enterprise, but particularly so for those who underwrite insurance coverage.

CLIMATE SURPRISES

In the climate change arena also, we are learning that the climate of a region sometimes changes abruptly and unpredictably. By its definition, climate is a long-term outcome of the day-to-day variability in temperature, precipitation, cloud cover, wind speed, and other factors. Therefore, by abrupt climate change we refer to major changes taking place over a decade or two, or even a century. Much of the attention given to recognizing and understanding the causes of abrupt climate change has been focused on the oceans, and the way that ocean currents move heat around the globe. Most geography courses, whether in elementary school or in a university, will point out that western Europe and Scandinavia are much milder than one might expect considering their rather northerly location, at about the same latitude as Hudson Bay in Canada. This occurs because an ocean current, the Gulf Stream, carries water warmed in tropical latitudes to the far northern reaches of the Atlantic Ocean, thereby warming Iceland and the adjacent European countries.

Geological evidence has identified times in the past when this current appears to have been interrupted, when the global conveyor belt of heat has slowed or stopped abruptly. A preliminary and rather tentative explanation of this phenomenon can be visualized somewhat like an escalator in a department store. The moving stairs of the escalator carry shoppers up to the floor above, but then the stairs disappear and return to the lower floor. The stairs comprise a large loop, with part of the loop visible, and the other part traveling the opposite direction out of sight beneath us. So it is with the ocean currents in the Atlantic. The Gulf Stream is moving northward along the surface, but there is a counter current moving southward along the ocean floor. The northbound surface current is a warm current from the south carrying heat to high latitudes; the southbound current is a cold current that carries cold and dense Arctic waters all the way to Antarctica along the ocean bottom. In both the department store escalator and the ocean, if one part of the loop encounters an obstacle, the entire loop shuts down.

So what is it that might bring the ocean current to a halt? First we must have an idea of what factors are important to maintaining the oceanic current loop. An essential component of the engine that drives the ocean current is the formation of the cold and dense bottom water that runs southward from the far north. Two factors influence the formation of such water: its temperature, and its salinity. Warm water is less dense than cold water and, therefore, more buoyant; generally, it will form near the surface of the ocean, where the sun's heat is absorbed, and travel in surface currents. Cold water is denser, however, and will generally travel along the ocean bottom.

The salinity also plays an interesting role in the density of water. The saltier ocean water is, the denser it is. Conversely, the less saline the water, the more buoyant it is. Therefore, fresh water coming into the oceans from rivers tends to stay near the surface, until it gets well mixed with normally salty ocean water. However, the annual formation of winter sea ice in the Arctic Ocean excludes the salt, which sinks downward as dense plumes of brine. In the far north Atlantic, the formation of dense bottom water is promoted by the very cold and salty water exiting into the Atlantic from the Arctic Ocean basin. This pathway is the only significant outlet of the Arctic Ocean. The dense bottom water then pushes southward, thus completing the loop of the oceanic conveyor belt. The combined effects of both temperature and salinity on ocean currents have given rise to the technical term thermohaline circulation.

The crucial link in creating and maintaining the current loop is the process of 'bottom water' formation in the north Atlantic. How might the current suddenly stop? If the surface water becomes less dense, it will 'float' rather than sink to the bottom. This could happen through a gradual warming of the Arctic Ocean and surrounding areas. The ocean water, with less ice cover, would be warmed directly by the sun. At the same time, it would become less salty, as the permafrost of the high Arctic melted, sending increased volumes of fresh water into the Arctic Ocean through the great north-flowing rivers, the Ob, Lena, and Yenisei that drain vast regions of Asia, and the Mackenzie flowing to the Arctic through Canada. With the area of annual sea ice diminishing, fewer brine plumes would form. At some point in this process, the water would become too buoyant to sink, and the bottom water would lose its replenishment. With the conveyor stalled at this important turning point, the transport of heat northward by the Gulf Stream would cease.

It is ironic that such a warming of the Arctic might lead to a shutting down of the oceanic conveyor belt of heat and unleash a severe chill across Western Europe. Ominous signs are already appearing: the summer of 2000 witnessed large expanses of open water at the North Pole and an unimpeded transit of the Northwest Passage by a Canadian icebreaker, without encountering ice. The summer of 2001 saw another passage through this far northern waterway, a record-breaking run from Greenland to Alaska in just seventeen days by a trawler without an ice-hardened hull.[4] Submarine measurements of the Arctic Ocean ice pack show a thinning of some 40% over the last half of the twentieth century, and satellite photos reveal that the area of the Arctic Ocean covered by sea ice has been diminishing over that same time. Whether or when these warming trends in the Arctic will interrupt the present pattern of ocean circulation simply cannot be predicted with any degree of certainty. Already ocean bottom measurements of the deep current at one location between Iceland and Europe have shown that in 2001 it was slower by half than it was four decades earlier.

Once before, some 11,000 years ago, the thermohaline circulation of the North Atlantic was substantially altered. That occurrence was likely caused by the delivery of a very large volume of cold fresh water into the North Atlantic, when the North American ice sheet had retreated northward to a position where the glacial meltwater was able to use the St. Lawrence River valley to reach the sea, instead of flowing southward into the Gulf of Mexico via the Mississippi River. The cooling and freshening of the Gulf Stream was fairly rapid, taking place over only a few decades, and lasted for about a thousand years. The temperature in Western Europe dropped some $5\,°C$ ($9\,°F$).

This is somewhat analogous to walking in the mountains in the dark but not realizing that you are approaching the edge of a cliff. Everything seems normal until you step off the cliff. Similarly, when you are popping corn, heat slowly warms the kernels until they reach a

---

[4] *Chicago Tribune Magazine*, 30 September 2001.

temperature threshold, at which time there is suddenly a lot of action as the kernels burst. Or milk chills in a bottle unnoticed until it freezes and expands, either popping the top off or breaking the bottle. In each of the latter two examples, with sufficient observation of the system, we could probably recognize that we were approaching the threshold of major change. But we do not yet have sufficient knowledge about the behavior of ocean currents to know how close we are to the cliff edge of the thermohaline circulation of the ocean. Being aware that there is a cliff in the vicinity is, however, useful knowledge.

The consequences of this unpredictability in nature usually come as a big surprise. Natural systems may seem to be perking along in their ordinary mode of behavior, perhaps changing slowly within well-defined bounds, when suddenly the system lurches into a new and apparently stable behavior, goes into wild gyrations, or shuts down altogether. This type of behavior, the abrupt transition of a dynamic system from one state to another, has been studied extensively in a theoretical context through a branch of mathematics known as complex systems theory.

Does the concept of 'insurance' have relevance in situations like this? Clearly insurance here must be thought of in national and global terms. Moreover, estimating how much to pay for 'insurance', that is, setting the premium, is no simple undertaking. Can we even envision the full range of consequences that a change in oceanic circulation might bring? Can we identify regions and countries that would be most impacted? Should a 'premium' be uniformly assessed across all nations, or should there be adjustments in establishing premiums that reflect which countries will be most severely impacted? Or should a variability in premiums reflect which countries have contributed disproportionately to the problem? Such questions seem almost impossible to answer objectively, and many people have thrown up their hands in frustration in trying to come to grips with such issues.

We do have, however, an analogy that may give us some modest comfort as we confront the large topic of global climate change and its possible consequences. Nations of the world all determine how much of their national wealth they want to devote to national defense. In simple terms, armed forces comprise an insurance policy for nations. In principle, they provide protection against external aggression (and sometimes internal insurrection). When we pay our homeowner's insurance premiums we actually hope that the money has been wasted, because we never want a claimable event to take place. With armed forces, we too would prefer never to call them into action, but we often concede that they are a necessity, at least at some level.

It is impossible for most people to estimate what a proper insurance premium for protection against external threats should be, that is, to determine rationally how much to spend on national defense. Nevertheless, most countries of the world do have a military component in their annual budgets. With very few exceptions, countries of the world, large and small, wealthy and impoverished, invest in such insurance. The amount that any nation chooses to spend depends on many factors, some only quasi-rational, related to quantitative assessments of the strength of possible adversaries, to long-established patterns of corruption and corporate welfare, and to some clearly emotional issues such as national image and pride. However ultimately I would assert that the fundamental reason for countries of the world investing in national defense is that their citizens simply believe it is necessary for national security. Individuals probably cannot easily articulate any detailed rationale for the size of the armed forces, nor exactly what weaponry the forces should have. Those details are left to professional politicians, civil servants, generals, and admirals. But the fundamental judgment has been made: we need armed forces for protection. With their armed forces called into action in Afghanistan following the terrorist attack in New York, most Americans probably were not complaining about the costs of military preparedness.

Similarly, it is impossible for most ordinary people to assess rationally the premium that we as nations should pay in order to avoid, or to prepare for, global climatic surprises. But I have every confidence that once citizens understand that climate change is real and that it poses threats, there will be a commitment to spending some of the national treasure for remediation and adaptation measures that will cushion an uncertain future. Citizens will not dwell on detail anymore than they do with national defense. They will simply tell the government and industry to get on with developing a climate-change insurance policy, comprising remediation and adaptation in some proportion, and in full recognition that it is going to cost something.

AN OUNCE OF PREVENTION ...

In 1240, Henry de Bracton advised us "An ounce of prevention is worth a pound of cure". The time-honored wisdom of this simple statement is clear enough: it is usually a lot cheaper to prevent a problem than to deal later with its consequences. It has been far less expensive to spend money developing vaccines than to deal with the medical and social costs of polio, smallpox, yellow fever, mumps, typhoid, and diphtheria. It is far cheaper to construct buildings to withstand earthquakes than it is to address catastrophic urban collapse such as that which accompanied the large earthquake in January 2001 that leveled the Indian city of Bhuj and severely damaged Ahmedabad. That earthquake caused a loss of life estimated between 25,000 and 100,000 persons, and it created a half-million refugees. It would have been far better, and ultimately far cheaper, to have more safety measures built into the nuclear reactor design at Chernobyl than to deal with the consequences of the radiation release there in 1986.

However clear the logic, it does not always guide the way we spend public or private money. Medical research focuses more on curing cancer than on preventing it; medical training emphasizes restoring health more than maintaining it. Even the health insurers, although wrapped in the banner that proclaims they are HMOs (health maintenance organizations), have historically been reluctant to pay for preventive or diagnostic services such as annual general physical examinations, birth control, adequate prenatal care, prostate tests, and mammograms. Even though HMOs now are more progressive in recognizing the cost efficiencies of these measures, they still pay little attention to environmental health issues such as air quality, even though they are footing the bill for increasing pulmonary conditions and asthma, which are reaching epidemic proportions across the USA as a result of smog.

One important reason the big health insurers say so little about environmental health, even though they see (and to some extent pay for) the health consequences of environmental degradation, is that some of their biggest customers are from industries contributing to the environmental degradation. An HMO in my home state of Michigan thinks twice before pointing a finger at General Motors or Detroit Edison, for fear of losing a large employer's health care business.

From the point of view of an HMO, perhaps the decision not to weigh in on the side of environmental and health protection has its own supportive logic. Perhaps the HMO feels that the air quality problems (and their insurance payouts) will persist even if they put pressure on their local customers to clean up their act. After all, much of the air pollution in Michigan comes from coal-fired power plants and factories in Illinois and Wisconsin, and other upwind states even further to the west. Similarly, much of the air pollution that is generated in Michigan and Ohio gets passed on to our neighbors to the east in Pennsylvania, New York and New Jersey.

And now, with Asian industrial pollution beginning to be detected in the air over the western states of the USA, traveling all the way across the Pacific, we are seeing that the problem is truly global. A local HMO may simply believe that the problem is much bigger and well beyond their capability to lead the charge for environmental protection, while at the same time losing customers on the field of battle. Global air quality, they would probably argue, is a problem for national governments and international relations.

Politicians who must address such issues often fall back on scientific uncertainty as a reason for maintaining the status quo. Not surprisingly, they seldom have conducted any scientific research[5] on

---

[5] In the 107th Congress of the USA (2001–2003), only two members had advanced degrees in physics.

their own. As noted in Chapter 2, many acquire their scientific perspectives from the very industries alleged to be damaging the environment, and the message these industries provide is that the scientific evidence is much too uncertain, much too inconclusive, to make any decisions affecting the future. In this context, of course, scientific uncertainty is no different than the uncertainties that obscured the future of the Social Security system when it was being designed. Had our representatives and senators adopted the same attitude in the 1930s, we probably would not have a Social Security program in place today. The debate surely must have had much of the same flavor: "How can one structure a retirement plan for seven decades into the future when we don't even know how many people will be living then? Or what their lifespan will be?" Had these questions, legitimate as they are, stymied the development and implementation of the Social Security system then, or at any time that such a system might be proposed, we would never have such a program. Only because the nation was ready to make decisions in the face of uncertainty do we now have a functioning system. Of course, the system today does not have exactly the same configuration as it did when it emerged from Congress in the 1930s. Several course corrections have modified it to reflect the ever-changing times.

THE PRECAUTIONARY PRINCIPLE: BETTER SAFE THAN SORRY

Richard E. Benedick, in his book discussing the ozone problem, wrote:[6]

> When we build a bridge, we build it to withstand much stronger pressure than it is ever likely to confront. And yet, when it comes to protecting the global atmosphere, where the stakes are so much higher, the attitude [of some people] seems to be equivalent to demanding certainty that the bridge will collapse as a justification

[6] *Ozone Diplomacy: New Directions in Safeguarding the Planet*, Harvard University Press, Boston, MA, 1991, p. 6.

for strengthening it. If we are to err in designing measures to protect the ozone layer, then let us, conscious of our responsibility to future generations, err on the side of caution.

You will of course recognize that taking the path of caution is exactly what this chapter describes in the discussions of emergency preparedness and insurance. Richard Benedick's remark about erring on the side of caution is an example of what has become known as the precautionary principle. In the context of climate change, it was enunciated at the United Nations Conference on Environment and Development (the 'Earth Summit') held in Rio de Janeiro in 1992. One product of that international conference was the Framework Convention on Climate Change, in which Article 3 states that the 160 signatory nations should "take precautionary measures to anticipate, prevent or minimize the causes of climate change and mitigate its adverse effects. Where there are threats of serious or irreversible damage, lack of full scientific certainty should not be used as a reason for postponing such measures ... "

Christine Todd Whitman as Governor of New Jersey (prior to being appointed to head the Environmental Protection Agency by George W. Bush) developed the precautionary concept further. In an address to the US National Academy of Sciences,[7] she said: "I believe policymakers need to take a precautionary approach to environmental protection. By this I mean we must 1) acknowledge that uncertainty is inherent in managing natural resources, 2) recognize that it is usually easier to prevent environmental damage than to repair it later, and 3) shift the burden of proof away from those advocating protection toward those proposing an action that may be harmful." Her third point argued that for too long the burden fell to environmentalists to show that the consequences of some action would be harmful; now she felt that the burden must shift to industrialists to make a persuasive case prior to the action that no harm would result.

---

[7]Christine Todd Whitman, "Effective policy making: the role of good science", an address delivered on 13 October 2000 in Washington, DC.

Moreover, she recognized that science operates on a fundamentally different time scale than does policy-making. Following an outbreak of a toxic microbe in 1997, fishing in some waters was prohibited to prevent possibly contaminated seafood from entering the human food chain. Scientists could not provide instant answers as to how the toxicity developed and whether contamination of seafood was likely, and the protection of public health could not wait indefinitely. The decision to ban fishing in certain areas was made on the best information at hand. "The absence of certainty is not an excuse to do nothing", Governor Whitman said, adding later, "If we want good science we cannot rush it."

Scientific studies later showed that the toxic microbe did not infect the marine life, but the Governor defended the fishing ban, stating, "I believe it was prudent to err on the side of public health", a clear declaration of the precautionary principle. She acknowledged there had been an economic downside to the decision but pointed out that, had there been contamination and illness, the deleterious economic impact on the seafood and tourist industries would likely have been even more severe.

Science can never produce complete certainty, and definitely not on a schedule. As science writer David Appell notes:[8] "...for the enormously complex and serious problems that now face the world – global warming, loss of biodiversity, toxins in the environment – science doesn't have all the answers, and traditional risk assessment and management may not be up to the job. Indeed, given the scope of such problems, they may *never* be." This perspective was enunciated in 1999 by the American Geophysical Union (AGU), the premier professional society of climate scientists, in a position paper[9] on climate change and greenhouse gases: "In view of the complexity of the Earth climate system, uncertainty in its description and in the prediction of changes will never be completely eliminated." The position paper

[8] *Scientific American*, January, 2001, p. 18.
[9] *EOS Transactions of the American Geophysical Union*, vol. 80, n. 5, 1999.

ended with the statement "AGU believes that the present level of scientific uncertainty does not justify inaction in the mitigation of human-induced climate change and/or the adaptation to it."

We should be careful not to let incomplete knowledge be used to stall the taking of precautionary steps. That we do not know everything about a complex system does not mean that we know nothing. The mantra "Take no actions until we have more answers" is usually a thinly disguised plea for the *status quo* by those with vested interests in maintaining it. More research does not even necessarily mean less uncertainty – as we learn more about a system, it may prove to be even more complex than we imagined, and have even greater uncertainty than first thought. Like it or not, we must make decisions in a timely fashion on the basis of information available. While it is true that sometimes haste does make waste, it is also true that doing nothing is often worse. Asking for certainty about the future before taking any steps to address today's problems is a bad formula. If steps we take today later prove to have flaws, course corrections can then be applied.

12 A climate of uncertainty

If we begin with certainties, we shall end in doubts, but if we begin with doubts, and are patient in them, we shall end in certainties.

Francis Bacon

We have walked a long way through the garden of uncertainty and have seen a multitude of flowers and a few weeds, much elegance and a little untidiness. The garden is not a formal garden, laid out geometrically, tended immaculately. It is a garden with many hidden recesses, in places a maze full of surprises, with each plot revealing something not seen before. As we near the end of our tour, we have come to recognize that uncertainty, just like a flower, can be found in many places and presents itself in different colors and intensities.

Throughout this walk through the garden of uncertainty, we encountered many aspects of global climate change: taking Earth's temperature, local trends and global averages, flood probabilities, ozone depletion, science education, industrial propaganda and obfuscation, media confusion, reconstructions of past climate, computer models of the week's weather and the century's climate, and insurance for an uncertain future. In each domain of the garden, the tie to climate change was bundled into a discussion of other natural phenomena and human activity, with uncertainties that paralleled or shared characteristics with the uncertainties of climate change. In this final chapter, I will pull together these components of climate change and address the attendant uncertainties cohesively, as a representation of both the struggles and achievements of climate science, and of the hills yet to climb.

In the last decade of the twentieth century, probably no other scientific topic saw more widespread discussion than global climate change, with a special focus on the temperature at the Earth's surface. The discussion of this important and fascinating subject can be framed around a series of questions, each of which can be posed and addressed independently but when taken together embrace the entire subject of global climate change rather comprehensively. In particular these questions and their answers illustrate the spectrum and range of uncertainty in our understanding of the global climate system, and the challenges and opportunities that the uncertainty presents. Each question carries uncertainties of different character and magnitude, of different color and intensity. Let me now pose the questions and proceed to discuss each one.

1. Has Earth been warming over the past century?
2. If so, what is causing the warming?
3. What will be the consequences of this change?
4. What can be done to remediate or accommodate the change?

HAS EARTH BEEN WARMING OVER THE PAST CENTURY?
The fundamental body of evidence that has been consulted to answer this question has been what we call the 'instrumental record'. The instrumental record essentially comprises a very great number of temperature measurements acquired from thermometers. These measurements are taken at meteorological observatories distributed unevenly across the continents, on islands in the oceans, on anchored and floating marine buoys, and by many ships at sea. In Chapter 5, we considered many of the problems that need to be addressed in making these individual measurements and aggregating them into a global average temperature for a given year. It is this record, the direct measurement of temperature at many places, which reveals that the average temperature of the atmosphere just above the Earth's surface has increased by about $0.6\,°C$ ($1.1\,°F$) over the twentieth century. The uncertainty in this number is expressed by a range of temperature in which the increase very likely resides. With 95% probability that range is 0.4–$0.8\,°C$.

We gain added confidence about this result because virtually the same result has been determined independently by three different research groups, at the National Climatic Data Center and at NASA in the USA, and at the Climate Research Group of the University of East

Anglia in England. Each group has established its own quality control measures about how to adjust for changes in instrumentation, how to average individual temperature measurements in order to achieve a global average, and how to filter out the effects of urbanization. Towns and cities tend to create 'heat islands', because of the ability of buildings and paved streets to absorb and retain heat, and thereby push the temperature upward from what might be observed at that location were it not developed. Early in the analysis of weather station records, it was recognized that urbanization might impart an upward bias to the instrumental record. Might not the apparent 'global' warming be indicative of nothing more than the fact that many weather stations are located in or near cities?

Fortunately, it is easy to address and dismiss this doubt. First one must remember that Earth's surface comprises both continents and oceans, representing approximately 30% and 70% of the surface, respectively. Clearly urbanization is not an oceanic phenomenon, yet the sea surface temperatures alone show an increase over the twentieth century of roughly the same amount as has been estimated for the globe as a whole. Second, there is a great amount of redundancy in the data from the weather stations on the continents. On any given day, there may be more than 5,000 stations reporting temperature readings, some in or near cities, some in rural areas. Analyses have been done in which all data from urban stations were excluded; only data from rural stations were used. Even though there were only around 2,000 stations that qualified as being unambiguously rural, the analyses show no essential differences between rural-only estimates of the century-long trend and estimates derived from all of the available data. This provides a clear confirmation that the corrections that have been made to the urban records have effectively removed the 'heat island' effect. One can confidently conclude that the warming of the globe is not some geographic artifact associated with where the temperature measurements were taken.

However, long after the legitimate questions about the urbanization effect had been satisfactorily addressed, there continues to be a recurrent chorus of doubt voiced about whether the instrumental record of global warming has been properly purged of the urbanization effect. Many of the voices in this chorus are those of the fossil fuel industries, who go to great effort to avoid implicating the combustion of coal and oil as a likely principal cause of the warming. Their strategy includes denying that global warming is taking place, by maintaining that the instrumental record is contaminated by the urbanization effect.

Another effort to raise uncertainty about the surface instrumental record centered on measurements of the average temperature of the lower five miles (eight kilometers) of the atmosphere determined both from weather balloons and by satellites orbiting more than a hundred miles above the surface. Over the past two decades, these measurements have shown that in this region of the atmosphere, well above the surface, there has been significantly less warming than observed at the surface. Much has been made of this difference to suggest that the surface measurements were in error. Responsible scientists placed these differences in perspective, noting that yet higher in the atmosphere the temperatures were even cooling as a consequence of ozone depletion. They also pointed out that the observations of what the middle and upper atmosphere were experiencing did not negate the observations of what was happening at the surface. These independent observations, all likely correct, simply point out the need for a better understanding of what processes control the vertical temperature profile of the atmosphere.

The conclusion that Earth's surface is warming, of course, does not stand or fall solely on the instrumental record. In addition to the direct instrumental evidence of warming, scientists have also observed a broad array of changes that arise as a consequence of warming. Most impressively, ice is melting over most of the globe. In the high Arctic, the ocean surface is frozen into sea ice. The extent of the ice varies seasonally, but over the past half-century the annual average area has shrunk by more than 10%. Moreover, in that same time interval, the thickness of the sea ice has diminished even more dramatically, by some 40%.[1] There is a very real possibility that, if warming continues at its present rate, the Arctic Ocean could experience ice-free summers sometime in the present century.

The Antarctic is also showing major changes. Enormous blocks of the floating ice shelves on the periphery of the continent have been breaking off with unprecedented frequency. Since 1995, the Larsen Ice Shelf along the east side of the Antarctic Peninsula, the Filchner–Ronne Shelf in the Weddell Sea, and the Ross Ice Shelf on the west side of the continent have all lost dramatic pieces of ice. The sizes are staggering: in 2002 the Larsen shelf lost an area the size of Rhode Island, and the Ross shelf disgorged a piece the size of Delaware. Even larger tracts of ice split from the Ronne shelf in 1998 and the Ross Sea in 2000. The massive fragment set loose in the Ross Sea was about one-third the size of Switzerland! Not all of Antarctica, however, appears to be affected in the same way. Some small shelves on the east side of the continent are stable, or even growing slightly, and the dramatic warming seen along the Antarctic peninsula is not characteristic of the temperature record in some other regions of the continent.

In more temperate regions, mountain glaciers are melting back to higher levels in the valleys they occupy. This melt-back is evident virtually every place with mountain glaciers: in the Alps of Europe, in the Andes of South America, in New Zealand, in East Africa, in the Himalayas, in Greenland, in the Rocky Mountains of North America. There are abundant historical observations and photographs of the former extent of many glaciers that show how much melting has taken place in the past century. Just as with the Arctic summer sea-ice, many mountain glaciers will probably disappear within the present century if present warming rates continue. The fabled 'snows of Kilimanjaro', already greatly diminished, may not last another two decades,[2] and Glacier National Park in the USA may have no glaciers in the second half of this century. As the warming creeps up the mountainsides,

[1] Rothrock, D. A., Yu, Y., and Maykut, G. A., *Geophysical Research Letters* vol. 26, pp. 3469–3472, 1999.
[2] *New York Times*, 19 February 2001.

life forms acclimated to cool temperatures are forced to move higher. Following in their footsteps are other plants and animals taking advantage of the expanding domain of warmer temperatures; already malaria-bearing mosquitoes are invading the foothills of mountainous regions previously malaria-free.[3]

Studies of the times of freezing and melting of lakes and rivers in the northern hemisphere each year have revealed a long-term trend toward later freezing and earlier melting, cutting the 'ice-season' by about twelve days over the past century.[4] In complementary fashion, the growing season in Alaska has expanded by ten days over the past thirty years.[5] The southern margin of snowfall in eastern North America has gradually moved northward by some twelve miles (twenty kilometers) since 1970 and the number of days each year with temperatures below freezing is diminishing. In the UK, flowers are blooming and birds are laying their eggs earlier in the spring. All of these phenomena point to decades or more of sustained warming.

Sea level has also risen about four inches (ten centimeters) in the past century. Some of this rise results from water from the melting of ice on the continents eventually making its way to the sea, but more than two-thirds of the change in sea level comes from simple thermal expansion of seawater. Many substances expand when heated, and water is no exception. A painstaking assembly of seawater temperature data, taken at various depths up to 3000 meters in all of the world's oceans, indicates that the heat content of the oceans has increased significantly over the past fifty years.[6] It is not just the surface that has been warming, but indeed the entire water mass of the global ocean. The additional heat content of the oceans is fully consistent with the observed sea-level changes thought to be caused by thermal expansion of seawater.

[3] Epstein, P. R., *Scientific American*, pp. 50–57, August 2000.
[4] Magnuson, J. J. et al., *Science*, vol. 289, pp. 1743–1746, 2000.
[5] Running, S. W. et al., *EOS Transactions American Geophysical Union*, vol. 80, n. 19, 1999.
[6] Levitus, S. et al., *Science*, vol. 292, pp. 267–270, 2001.

Similar evidence comes from the rocks beneath the surface on the continents. My own geophysical research has for many years involved taking the temperature of the Earth in boreholes drilled into the crust. These measurements, in more than 700 boreholes on all the continents, show that the upper few hundred meters of rock over much of the globe have been warming over the past several centuries. The warming in the twentieth century found in the rock temperatures is fully consistent with the warming of the surface revealed by the instrumental record.

Year by year, little by little, the warming of the planet – its surface, its waters, its rocky crust – has become apparent in myriad ways. It is precisely because we do not rely on a single type of observation, nor on the data of a single research team, nor on observations from a limited geographic region, that we are so confident that global warming is real. One can no doubt find questions to raise about this instrument, or that data assemblage, or someone's statistical analysis. But when so many lines of evidence independently tell the same story, objective people find the story compelling and the conclusions inescapable. And so, at the beginning of the twenty-first century, the debate about whether Earth is warming is over. Mainstream atmospheric scientists now assess the probability that the warming is real, as opposed to some inexplicable misreading of the observations, at better than 99%. In other words, it is a 'virtual certainty' that the warming is real. Indeed, most skeptics have now conceded that Earth has warmed in the twentieth century. The occasional insistence that this remains an open question comes only from the climatological equivalents of the Flat Earth Society, or those whose economic or political agendas motivate them to distort or deny the plethora of evidence.

So, with the answer to this first question clearly in the affirmative, let us move on to the next question.

WHAT ARE THE CAUSES OF THE CHANGE?
Virtually all climate scientists will acknowledge that Earth's climate at any given time is the product of several factors. These include the

amount of energy received from the Sun, how that energy is reflected or absorbed by the Earth, and how the planet loses energy back to space. The way the climate system works in general is that our planet receives a certain amount of radiative energy from the Sun, governed both by the radiance of the Sun and our distance away from it. Planets closer to the Sun receive more energy, and those further out in the solar system receive less. With the radiant energy diminishing with distance from the Sun, it generally gets colder the farther out in the solar system a planet is situated.

About 30% of the radiant energy that arrives at Earth is reflected from Earth's surface back to space, particularly by the white polar ice. The remaining 70% of the incoming solar energy is absorbed by the darker rocks and vegetation of the continents and the waters of the ocean, thus warming the Earth. But that cannot be the whole story, because if it were, we would get a little warmer everyday as the Sun continues to deliver radiant energy and the planet absorbs it. How does Earth avoid this?

It is well known, at least among physicists, that all objects radiate energy at a rate related to their temperature. An electric space heater radiates heat that can be felt at a distance, and a wall of a building bathed in sunshine all afternoon radiates heat well into the evening. Similarly the entire Earth, taking a perpetual sunbath, does not permanently retain that energy; it too radiates away the solar energy it receives. Over the eons of Earth history, the surface temperature of the planet has adjusted so that it is just adequate to radiate away the radiant energy it receives from the Sun. That temperature represents an equilibrium between incoming sunshine and outgoing 'Earthshine'.

But the climate system has many more complexities than portrayed by this simple equilibrium scenario. Most importantly, Earth's atmosphere places a significant overprint on the fundamental Earth–Sun interaction, via the greenhouse effect. The atmospheric greenhouse operates something like a live animal trap – it is easy to get in but hard to get out. It is easy for most of the inbound solar energy to pass through the atmosphere, but it is much more difficult for the energy leaving Earth to escape. This asymmetry arises because the incoming and outgoing energy occupy different parts of the electromagnetic spectrum. The incoming radiant energy from the Sun is principally in the 'visible' band of the spectrum (so-called because we can see it), whereas the outgoing radiation from Earth is mainly in the longer wavelength (and invisible) infrared part of the spectrum. The atmosphere is rather transparent to most of the incoming energy, but it traps some of the outbound radiation. This trapped energy warms Earth's surface to a higher temperature than it would have if the Earth had no atmosphere. The trapping mechanism is the absorption of infrared radiation by certain gases in the atmosphere, particularly by water vapor and carbon dioxide. Both of these gases are present in the atmosphere only in trace amounts, together making up a very small fraction of the atmosphere. But never underestimate the significance of very small quantities, whether they be greenhouse gas concentrations in the atmosphere or pesticides dissolved in groundwater – both can have profound effects.

The greenhouse effect is not some theoretical concept whose existence is open to debate. It is a real physical effect that can be observed. In terms of the language of probability mentioned in Chapter 6, the existence of the natural greenhouse is a 'virtual certainty'. Earth has had a greenhouse atmosphere almost from its birthday, and we should be thankful for the fact. Without a natural greenhouse, Earth's surface would be significantly colder, so much so that the oceans would be frozen solid and Earth would be a far less hospitable place. At the most fundamental level, the Sun and the atmospheric greenhouse collaborate to produce Earth's long-term average surface temperature.

But how might these players be contributing to the short-term changes that we have observed in the twentieth century? The answer to that question centers on whether the radiant energy from the Sun has *changed*, or whether the effectiveness of the greenhouse has *changed* in the time frame that we have observed the recent warming. The central question is not whether the Sun and the greenhouse affect climate – we know that both are very important. The key question is whether and by how much either of them may be changing from their long-term behavior. In recent years, important research programs have been mounted to determine the historical variability of both the Sun and the greenhouse, and to compare their variability with reconstructions of how Earth's surface temperature has varied in the same time interval. All of this research proceeds in the shadow of the uncertainty that characterizes reconstructions of the past, as discussed in Chapter 9.

Solar variability

It has been known for centuries that the Sun undergoes a cyclical change in its appearance; small dark spots appear, grow in number, and later disappear from the face of the Sun. The coming and going of these dark spots is a manifestation of a cyclical fluctuation in the energy output of the Sun. The entire rise and fall takes about eleven years and then is repeated. The number of dark spots have been routinely counted on a daily basis for the last twenty-two cycles, and on a less regular basis back some 400 years. During the last two sunspot cycles, roughly in the interval 1979–2001, Earth-orbiting satellites have been busy measuring how much the solar output varies throughout each cycle. This 'calibration' is then used to estimate the solar radiance in times past on the basis of the historical record of the sunspot numbers.

To estimate fluctuations in solar radiance prior to the systematic counting of sunspots, we resort to other means. The production of certain chemical isotopes, such as carbon-14 and beryllium-10, is influenced by energetic particles ejected from the Sun, and the variable concentrations of these elements in fossils and rocks has been used as a proxy, or substitute, for estimating fluctuations in solar radiance. These isotope methods enable scientists to estimate solar variability back several thousand years.

Changes in the greenhouse

Just as with the investigations of solar variability, the atmospheric concentrations of several greenhouse gases have over the past forty years been measured directly with instruments at many places in the world. This type of measurement is a rather straightforward and routine operation, not particularly difficult nor controversial. These observations, begun in the late 1950s as part of the International Geophysical Year program, have shown a very steady increase in greenhouse gas concentrations in the atmosphere throughout the entire period of observation. In just the forty years of observation, the concentration of carbon dioxide has increased by 20% over its mid-twentieth century level. Once again, using probabilistic language, the reality of this change in the greenhouse is a 'virtual certainty'.

The record of greenhouse gas concentrations, however, goes back much further in time than just the most recent half-century of direct measurements. Scientists have been able to determine these concentrations for many thousands of years into the past, thanks to the existence of an icy archive of trapped bubbles (mentioned in Chapter 9) that are atmospheric 'fossils', samples of the atmosphere at the time they were trapped. The bubbles show that in 1750, around the beginning of the industrial revolution, the carbon dioxide level in the atmosphere was about 280 ppm, well below the 380 ppm observed at the end of the twentieth century. Today in the early years of the twenty-first century, the carbon dioxide continues to climb an additional half-percent each year over the pre-industrial level. If that rate of growth continues, and there is little to suggest that it will not, carbon dioxide in the atmosphere will reach double its pre-industrial level, that is a full 100% increase, around the year 2070.

The thick ice of Antarctica preserves a record of atmospheric carbon dioxide over the past 420,000 years that lets us place the more recent changes in a really long-term perspective. The Antarctic ice cores show that the present-day level of carbon dioxide exceeds the concentration seen at any time in the long Antarctic record, a period of almost a half-million years. Moreover, as mentioned in Chapter 9, the temperatures over this long period of time (as determined from the isotopic chemistry of the ice) correlate extremely well with the variations in carbon dioxide: high temperatures with high carbon dioxide, low temperatures with low carbon dioxide. If that strong correlation continues, the present level and rate of growth of carbon dioxide makes higher temperatures later in this century an inescapable outcome.

Carbon dioxide is not the only greenhouse gas for which concentration is increasing; methane has already doubled its pre-industrial level, and the CFCs, the chemicals that have been responsible for ozone depletion, have also played a significant greenhouse role in the twentieth century. Fortunately, for both the ozone and the greenhouse, the phase-out of CFC production resulting from the Montreal Protocol of 1987 will lead to the slow decline of these gases in the atmosphere.

Atmospheric particles

In addition to solar variability and changes in the greenhouse, other factors also play a role in the changing climate. One important factor that has accompanied the increase of greenhouse gases in the atmosphere is the increase of microscopic particles and droplets known as aerosols. Both the greenhouse gases and the aerosols arise from the combustion of fossil fuels and both have increased during the growth of industrialization. The aerosols, however, appear to offset part of the greenhouse warming, because they reflect incoming sunshine back to space before it can reach Earth's surface. Similarly, the occasional volcanic eruptions that we talked about in the previous chapter can deliver ash and dust to the atmosphere that block some sunshine for a year or two.

What drivers of change are significant?

Which of these drivers of climate change – a variable Sun, a strengthening greenhouse, atmospheric aerosols, volcanic eruptions – can we point to as the cause of the recent warming? The history of each has been reasonably well studied so we can examine their relationship to Earth's temperature in both pre-industrial and industrial times. In the pre-industrial period of human history, the greenhouse was steady at its natural level and the industrial aerosol load nil. We must look to the natural drivers, the Sun and the occasional volcanic eruption, to explain pre-industrial climatic fluctuations. The coincidence of the deepest chill of the Little Ice Age with a long interval of very few sunspots (1645–1715) strongly suggests a solar influence at that time. And the 'year without a summer' that followed the eruption of the volcano Tambora in 1815 demonstrated the significant albeit brief effect of volcanic aerosols and ash in the atmosphere.

However, the upward temperature trends of the twentieth century, particularly in the last half of the century, cannot be accomplished with just the natural drivers alone. In fact, computer simulations that employ only the natural drivers, solar variability and volcanic aerosols, actually show a slight cooling in the latter half of the twentieth century, owing principally to an upturn in volcanic activity. Simulations of the twentieth century that do not include the rapidly strengthening greenhouse and the accompanying industrial aerosols as drivers, fall increasingly below the actual temperatures that were observed. Only by adding in the strengthening greenhouse effect and industrial aerosols that result from the burning of fossil fuels do the simulations recreate reality and track the observed warming. It is on the basis of these climate simulations that the climate scientists on the Intergovernmental Panel for Climate Change (IPCC) concluded in 2001 that most of the warming observed over the last fifty years is attributable to human activity.

How much uncertainty accompanies this conclusion? The IPCC assesses the probability that the conclusion is correct lies in the 'very probable' range, 90–99% certain. While the observations of the increases of temperature and of greenhouse gases were both rated as 'virtual certainties', quantifying the links between the two in a complex computer simulation made the conclusion modestly less certain than were the individual observations standing alone. Only five years ago, the IPCC was more cautious, saying that the balance of evidence only *suggested* a discernible human influence on climate, with a probability of being true at 66–90%, in the 'likely' range. Studies in the intervening years have led to greater confidence and less uncertainty about the causes of the planetary warming.

WHAT WILL BE THE CONSEQUENCES OF THE CHANGE?
Looking ahead to the consequences of climate change, we leave behind the uncertainties inherent in reconstructing the past and move into the uncertain undertaking of predicting the future. As we have seen, the uncertainties of prediction arise sometimes from not fully understanding how a system works and at other times, even with an adequate model of a system, from being unable to predict how the factors important to the system will evolve over time. The example of the US Social Security system developed in Chapter 7 illustrates this latter type; the equations governing the system are relatively straightforward, but the demographic projections have needed revision over time.

Uncertainties aside, however, assessment of the consequences of global warming is sometimes pre-empted by another question: "Who cares?" If the consequences of warming are all good, then there is little reason for concern and little need for remedial action. It will come as no surprise that the question emanates from the ever smaller but still highly vocal band of global warming 'contras'. With the past century of warming now documented with virtual certainty, and with the past half-century of that warming very probably attributable to human activities, the skeptics have retreated to a third line of defense. They are now telling us not to worry, that global warming will be good for us. The argument goes something like this: warmer weather will extend the growing season everywhere, and increased carbon dioxide in the atmosphere will make everything grow better and faster. Add milder winters to those benefits, and everyone will be better off. It all sounds very appealing. Just beneath these superficialities, however, the complexities of the real world temper such perspectives considerably.

Enhanced plant growth is itself a mixed blessing, because it will not be selective; weeds will keep pace with (or exceed?) beneficial growth and will require more human endeavor and herbicides for control. And not all plant species will benefit uniformly, so there will be displacement as dominant species take control of the territory. Moreover, enhanced growth requires more than just carbon dioxide fertilizer. Other nutrients, particularly some nitrogen compounds and of course water, must be available to enable the production of greater biomass. To take advantage of the increased availability of atmospheric carbon dioxide will require increased application of other fertilizers, with attendant water-quality problems appearing in the lakes and rivers of the watershed. And given that one of the consequences of atmospheric warming will be a progressive desiccation of the soil in many places, the effects of carbon dioxide fertilization may be largely offset by the loss of soil moisture. In experiments with plant growth at elevated levels of carbon dioxide, with other factors controlled to be unchanging, biomass did increase, but the nutrient value did not; rather it seemed to be have been spread more thinly through the greater biomass.

In a northern Alaska forest, scientists have looked for evidence that tree growth had accelerated as a result of carbon dioxide fertilization. What they found was a forest on the decline because of the warming the region has experienced. The carbon dioxide fertilizer was there for the taking, but the higher temperatures in the neighborhood had dulled the forest's appetite. Moreover, if there is a decline in the principal tree species making up the forest, other opportunistic species will move in to occupy the terrain. A musical chairs scenario will be set in motion, with each species seeking to settle in a new place; however, as in musical chairs, some will discover that there may be no new place to settle. It is totally naive from an ecological perspective to think that, under conditions of increased carbon dioxide fertilization, everything will stay put and just grow faster.

Underlying all assessments of consequences is the estimated range of how much warming will occur. Computer models of warming over the twenty-first century yield a range of temperature increases of about 1.5–$5.5\,°C$ (2.7–$9.9\,°F$), quite a large range that reflects uncertainty. But uncertainty about what? At a very basic level, this range of uncertainty derives not so much from uncertainties in our scientific understanding of the global climate system, but rather from the uncertainties of how the global population will grow, and what types and amounts of energy Earth's inhabitants will use. These are the uncertainties about the social and economic future, the same types of uncertainty that characterize the future needs of the Social Security system that were discussed in Chapter 7.

There are many intricate computer models of global climate developed independently by universities and climate research laboratories around the world. Given the same set of demographic and economic assumptions, the climatological projections of the future from this wide array of climate models are generally very similar. That so many research groups have made independent judgments about how to represent the fundamental physics and chemistry of climate in their models, and yet reach very similar conclusions about how the climate system will evolve, should allow us to have some confidence in the conclusions. The greater uncertainties about our climatic future stem not from climate science but from social science and the vagaries of human behavior.

Unfortunately, it is already apparent that the consequences of global warming, some of which were mentioned earlier as indirect evidence that substantiates warming, are not all beneficial. The most widespread consequence, one that absolutely no one has argued would be beneficial, is the rise of sea level. Over the past century, the sea has risen by about four inches (ten centimetres), largely as a result of the warming and attendant thermal expansion of the ocean water. The range of warming projected with various population and developmental scenarios is easily translated into a range of possible sea level change over the twenty-first century: an additional eight to twenty-eight inches (twenty to seventy centimeters) above present-day sea level. But the real impact of a rise in sea level is expressed in the amount of territory that will become flooded. On a very gently sloping coastal plain such as the eastern seaboard and Gulf coast of the USA, or where the Amazon, the Ganges, and the Irawaddy Rivers meet the sea, a small rise of sea level results in a big shift of the coastline inland, inundating large areas. Much of south Florida faces inundation in most of the twenty-first century scenarios, and many populated islands in the Pacific face a similar fate. A rise in sea level will mean profound changes for all of the cities of the world adjacent to the sea: New York, Miami, New Orleans, Amsterdam, Copenhagen, Tokyo, Buenos Aires – the list is endless. Sea walls and dikes will be both exorbitantly expensive and ultimately ineffective. It is one thing to wall off a portion of a small country like the Netherlands, and quite another to build and maintain a dike around a continent!

Another consequence of the warming of the lower atmosphere is its increased capacity for water vapor. There is considerable debate about what this will mean for the climate system. If it translates into thicker and more widespread clouds, which reflect more incoming sunshine back into space, it might slow down or stabilize the warming. But water vapor is also a greenhouse gas, and increasing its presence in the atmosphere might augment the warming already underway from carbon dioxide and other greenhouse gases. How this ambivalent behavior of water vapor in the atmosphere will play out is one of the larger uncertainties in climate science, but because it could tilt either way the uncertainty falls squarely in the middle of the probability scale.

However, another aspect of the atmosphere's increased capacity for water vapor is the increased likelihood of severe precipitation events, defined somewhat arbitrarily as an event when more than two inches (five centimeters) of rain falls in a twenty-four hour period. A detailed examination of international weather archives for the past century[7] has shown an upward trend in the land area that

[7]Karl, T. R., Knight, R. W., and Plummer, N., Trends in high frequency climate variability in the twentieth century, *Nature* vol. 377, pp. 217–220, 1995.

has experienced a severe rainfall event. The year 2000 witnessed a foot of rain (thirty centimeters) falling on northeastern North Dakota in 12 hours, and fourteen inches (thirty-five centimeters) inundating north-central New Jersey in 24 hours, amounts greatly in excess of the previous record-holding precipitation events in these areas.

It is one thing to estimate with high probability that the global average temperature will increase, and that the water vapor capacity of the global atmosphere will increase, but it is quite another thing to project how the departures from average will be distributed around the globe. Any average increase is made up of regions above the average and other regions below the average. Consequently, there is a much greater uncertainty in forecasting changes on a regional scale, even though there is high confidence in estimates of the average change over the entire globe. Different computer models, while concurring on global-scale trends, may disagree on how those trends will play out regionally. In the Great Lakes region of North America, several computer models forecast a drop in the Great Lakes water levels, but a few forecast a rise.

In the previous chapter, I also discussed surprises in the climate system. These are changes that might take place abruptly, as when some important threshold is reached that alters the thermohaline circulation of the Atlantic Ocean, ending the transport of heat to northern Europe. Or when a sizeable fraction of the Greenland ice cap slides into the ocean, raising sea levels virtually instantaneously like dropping an ice cube into a glass of water. Climate scientists do not yet know how to identify the thresholds for such events, and the uncertainties remain formidable.

WHAT CAN BE DONE TO REMEDIATE OR ACCOMMODATE THE CHANGE?
Here we move into new terrain, a topography that is not exclusively scientific. Addressing the many aspects of how to deal with a changing climate quickly leads to a confluence of science, economics, government, religion, and more. Much can and will be written about how

these relevant constituencies encounter and navigate these hills and valleys and confront each other on this terrain in the coming decades.

There are, of course, powerful defenders of the *status quo*, industries and politicians who argue that we should be paying no attention whatsoever to climate change. This militia of global-warming opponents has been scattering leaflets everywhere over this terrain, urging anyone and everyone that the best course is 'business as usual', maintaining the *status quo*. The arguments are now familiar: global warming is only a theory, the predicted effects are highly uncertain, scientists are not unanimous, the data are in error, more research is needed. Using classical scare tactics, they predict dire economic consequences, telling the population that if measures are taken to address climate change they probably will be ineffective, and in the process the economy will suffer and jobs will be lost. Some try to make a case, as I have just described, that climate change is something to look forward to, not derail. However, behind the curtain of uncertainty that they have tried to create, one easily sees their pressing concern: they believe that many of the changes proposed to address a changing climate will have adverse economic consequences for them. The public, and particularly the media, must continue to recognize arguments stemming from obvious self-interest and weigh them appropriately.

There is actually much that can be achieved technologically to slow the buildup of greenhouse gases in the atmosphere. First among them is simple energy conservation. Benjamin Franklin's observation that a penny saved is a penny earned is still relevant. A kilowatt-hour of electricity not used is a kilowatt-hour that does not need to be produced or purchased. Residents of California responded to and helped to alleviate the electricity crisis of 2001 in the only short-term way available to them: conservation. The success of conservation measures surprised many, particularly in terms of how quickly the effects became apparent.

In an only slightly longer time frame, there are many ways to conserve energy through improving the efficiency of the vehicles we drive and the appliances we use. Efficient hybrid-electric vehicles that get up to sixty miles per gallon of gasoline (twenty-five kilometers per liter of petrol) and require no immobile battery charging are already selling well in the USA. High-efficiency furnaces and air-conditioners, refrigerators, washers, driers, and water heaters are being marketed successfully, in part on the basis of the energy cost savings that will accrue over their lifetimes. In addition, insulating buildings to higher standards has a high potential for conservation. These measures require no additional research and development; they are all products on the market today. They can be immediately exploited to reduce greenhouse gas emissions to the atmosphere. They can be profitable to the shareholders, and they offer abundant employment opportunities.

A second line of technological remedies is the development of energy alternatives to the fossil fuels, energy sources that do not produce greenhouse gases. Some, like hydropower and wind power, have been long known and locally utilized. Today, wind energy has seen a technological renaissance that has enabled its utilization for electrical generation in significant amounts. Nuclear energy does not pollute the atmosphere but has other environmental hurdles to overcome that have slowed its growth as a fossil fuel alternative. In the aftermath of the September 2001 terrorist attacks in the USA, a new awareness of modern society's vulnerability to terrorism has defined a further barrier that nuclear energy must overcome: the safe and secure sequestration of fissionable waste from nuclear power generation, and the security of power plants from internal sabotage and external attack. Geothermal energy is locally attractive where geological activity has placed hot rocks near the surface. Direct generation of electricity via solar cells and in hydrogen fuel cells holds considerable promise for the not-too-distant future.

A third technological approach to taming the greenhouse is the sequestration of greenhouse gases, in effect capturing and storing them before they escape to the atmosphere. A pilot project to return carbon dioxide to the subsurface (in a geological reservoir beneath the North Sea from which oil and gas have been extracted) has already demonstrated considerable promise. Another suggested possibility is the pumping of carbon dioxide to the bottom of the ocean, where it can be absorbed and stored in seawater. Ocean storage, however, has been criticized as potentially unstable, and possibly having deleterious effects on oceanic ecosystems. The utilization of carbon dioxide by photosynthetic plants can be viewed as yet another process whereby carbon dioxide is removed from the atmosphere to temporary storage in biomass. Although the return of the Garden of Eden, as promised by some as a benefit of carbon dioxide emissions, is unlikely, countering the widespread deforestation occurring in many developing countries with programs of reforestation elsewhere can be a part of the solution.

Change, however, is inevitable; indeed it is already underway. Earth today is responding to the changes initiated at the beginning of the industrial revolution. Like trying to turn a large battleship at sea, Earth will be slow to respond to changes we make over the next fifty years. There is an immense amount of inertia in both the natural and social systems of Earth. The human population of the planet, with all its attendant needs, will continue to grow throughout the first half of the twenty-first century, and maybe beyond that. The large heat capacity of the oceans guarantees that the warming already experienced will continue to influence global climate decades into the future. Reducing carbon dioxide emissions to slightly below 1990 levels, as required by the Kyoto Protocol on greenhouse gases, will only slow the growth of the greenhouse potency, not reduce it. And the residence time of greenhouse gases already in the atmosphere, even in the absence of any new additions, is measured in decades and centuries. Like it or not, humans on Earth must realistically anticipate change from the impacts of the enhanced greenhouse already in place owing to the use of fossil fuels and the growth of population throughout the twentieth century.

Adaptation to change, therefore, is one avenue toward the changing future. What can be done? Let me point out just a few areas that will surely command attention. In the face of rising sea level, coastal land-use traditions will come under increasing stress, and nations will face important decisions about how much of the public treasury to invest in protective measures. Flooding of low-lying areas will force the relocation of substantial numbers of people, in ways similar to refugees dislocated by war. Public health activity will have to adapt to a greater range of tropical disease and heat-related health problems. Water will emerge as the premier resource of the twenty-first century, with competition for it surpassing the intensity of the twentieth century competition for energy. Indeed, the need for fresh water may be what ultimately makes solar energy economically competitive, as the principal energy source that will enable the large-scale desalinization of ocean water. The already-accelerating price of fresh water[8] will eclipse the increasing cost of energy necessary to acquire and distribute fresh water. And as the geography of agriculture changes in response to the changing patterns of precipitation and soil moisture that accompany the inevitable warming, national and international water policy will move to the forefront and center in the forums of political and economic debate.

What roles can the governments of the world play, on the one hand to slow the immense forces of human activity that are driving the climate of the planet into new and uncharted terrain, and on the other to prepare Earth's human inhabitants for the changes that are inevitable? Within the bounds of individual nations, there are governmental actions that can and do influence greenhouse gas emissions. Governments have at their disposal a wide array of tax policies that can provide incentives for alternative energy development and disincentives for the continued heavy reliance on fossil fuels. The tax levied on gasoline at the pump is directly reflected in the fuel efficiency of vehicles; the higher the cost of gasoline, the greater the public desire for more fuel-efficient vehicles. A direct tax on gasoline consumption has on occasion been criticized as being socially regressive, impacting the less-affluent segments of the population more severely. Such criticisms, however, can be addressed and redressed with adjustments elsewhere in the tax code. Governments have the power to set standards

---

[8] Already a liter of bottled water costs as much as a liter of gasoline in many places around the world.

for vehicle fuel efficiency but in the USA have been reluctant to exercise it in a meaningful way. In 2002, the US Senate again had a chance to impose higher efficiency standards for vehicles but, succumbing to pressure from senators in the auto-producing states, declined to do so.

There are also opportunities for collective action by the nations of the world. Although cooperative international steps to address global environmental issues is a field in its infancy, already there is one prominent success story: the Montreal Protocol, an international agreement negotiated in 1987 to phase out CFC production and bring an end to the destruction of stratospheric ozone. Only fifteen years after the agreement, CFC levels in the atmosphere have stabilized, and they will soon begin a decline. Although the ozone hole still persists, scientists anticipate the gradual recovery of the ozone as the CFC concentrations fall. If all goes according to script, the ozone distribution will be restored to its pre-CFC conditions mid-way through the twenty-first century.

Because the CFCs are themselves greenhouse gases, their elimination comprises a small step in reducing the potency of the atmospheric greenhouse. The Kyoto Protocol, a similar international agreement negotiated in 1997 to begin the reduction of carbon dioxide emissions, has gotten off to a slow and wobbly start because of the unwillingness of the USA to participate. Another truly international subject with broad implications for Earth's climate is the human population of the globe. We should never forget that total energy consumption is the product of per capita energy usage multiplied by the number of people. Stabilization of the population, locally and globally, will be one of the most effective pathways to slowing the growth of energy consumption. It is a topic that has never been on the table where climate change issues are debated.

危 机

These symbols are the Chinese word for crisis, which is made up of two words – danger and opportunity. The changes being wrought by humans on Earth's climate can be thought of as a crisis, both in the English language sense of a predicament, and in the Chinese sense of risk and opportunity.

The *perils* of global warming, half of the Chinese word for crisis, are today widely discussed and reasonably well understood. Much of the world has reached a consensus about the nature of the predicament: humans are altering the environment in which they live. However, the *opportunities* presented by global warming, the other half of 'crisis', are less well appreciated. Industries that have been motivated to examine their operations in terms of environmental consequences have discovered that there are large savings to be had in energy, water, and waste-handling costs. There are opportunities for advances in sustainable energy that promise substantial rewards for those that develop them. Environmentally benign modes of transportation will transform the twenty-first century in ways similar to the transformations brought about by affordable automobiles in the twentieth century. And some visionary industries already know that environment-friendly operations and products provide them with strong competitive advantage.

It is my impression that a more prudent and sanguine outlook on climate change is slowly developing, in a few industries, in some governments, and more widely in the general public. This outlook recognizes but is not deterred by the uncertainties associated with climate change. This emergent perspective recognizes that global climate change comprises both risks and opportunities, and that mitigation and remediation of climate change will involve both technological and social innovation. To be sure, from time to time there are setbacks to this emerging consensus, such as the policies of the George W. Bush administration in the USA with regard to energy policy, greenhouse gas emissions, and international strategies for addressing climate change. But the realities of global warming will press inexorably forward and eventually will force some decision-makers out of the ideological and political ruts in which they are stuck.

It is sometimes said that one is either an agent of change or a victim of it. While we are all 'victims' of the consequences of a warming climate, some of the greatest consequences hit those least able to be an agent of remediation: those populations dislocated by rising sea level. And among those that have the capability of introducing significant change are many that refuse to recognize a changing climate or acknowledge any responsibility for it. Ultimately they too will be victims, in a special way. They are so busy defending the *status quo* that they will fail to see the opportunities presented by this big inadvertent human experiment being carried out on the environment of life. Their perceived need to hold on to the present, and their fear of the uncertainties of the future, will preoccupy and distract them from taking advantage of the opportunities that the uncertainty offers. They are the ones that see the glass as being half empty, and their fear that it will soon be totally empty places them in a continually defensive posture. Jawaharlal Nehru, commenting about political timidity but appropriate here as well, remarked "The policy of being too cautious is the greatest risk of all."

People that see the glass as half full will be the successful agents of change. They will recognize an opportunity to fill the glass. They will be the ones whom uncertainty will stimulate rather than intimidate. They will be the ones who predict the future by creating it.

Index